The Guard
Of The E

How have volcanoes shaped our planet ?

©Patrick Donnet, 2023

All rights reserved. No part of this book may be reproduced or transmitted in any form
transmitted in any form, electronic or mechanical, including photocopying
photocopying, recording or any information storage system,
without the written permission of the author, except for a brief quotation in a review
or press article.

Contents

Introduction 5

Definition of a volcano 5
History of Volcanology 6
Importance of Volcanoes in Earth's History 7
Distribution of Volcanoes Worldwide 8

Basics of Volcanology and Volcano Formation 11

Structure of the Earth 11
Plate Tectonics 12
Formation of Hotspots 14
Magmatism and Volcano Formation 15
Igneous Rocks and Minerals 16
Composition of Volcanic Rocks 17
Magma and Lava: Differences and Properties 19

Types of Volcanoes on Earth and Their Characteristics 21

Shield Volcanoes 21
Stratovolcanoes 22
Scoria Cone Volcanoes 24
Submarine Volcanoes 25
Rift Volcanoes and Fissure Eruptions 27
Mud Volcanoes 28

Extraterrestrial Volcanoes 31

Volcanism on the Moon 31
Volcanism on Mars 32
Volcanism on Io 34
Other Examples of Volcanism in Our Solar System 35

Volcanic Forms and Structures...37

Volcanic Cones 37
Lava Domes...38
Lava Flows 39
Basalt Plateaus 40
Evolution of a Volcano over Time 41

Volcanic Products 43

Types of Lava 43
Lava Tubes...44
Tephra and Volcanic Ash...45
Volcanic Gases 47
Lahars and Mudflows..48
Pyroclastic Flows and Ejections 49

Processes of Volcanic Eruptions 51

Effusive Eruptions 51
Explosive Eruptions...52
Fumaroles and Solfatara 53
Volcanic Tsunamis...54
Factors Influencing Eruption Types 55

Famous Volcanoes and Historical Eruptions............................. 57

Mount Vesuvius and Pompeii.. 57
Mount Pelée and Saint-Pierre...58
Krakatoa and its Global Effects 59
Mount St. Helens and Modern Awareness... 61
Other Famous Volcanoes..62

Monitoring and Forecasting Volcanic Eruptions 65

Tools and Techniques for Volcanic Monitoring... 65
Precursor Signs 66
Eruption Modeling and Forecasting Methods 67
Limitations and Challenges in Prediction...69

Volcanic Risk Management 71

Mapping of Volcanic Hazard Zones 71
Urban Planning.. 72
Education and Awareness 73
Alert Systems and Evacuation Plans 75
Crisis Management 76

The Impact of Volcanoes on the Environment and Climate.... 78

Geological and Geomorphological Impacts.. 78
The Impact of Volcanic Gas Emissions 79
Volcanic Aerosols and Climate Cooling... 80
Volcanoes and the Carbon Cycle.. 81

Volcanoes and Life 83

Ecosystems around volcanoes 83
Organisms adapted to volcanic environments................................. 84
The role of volcanoes in species diversification............................... 85
Volcanoes and the origin of life... 86

Volcanoes in Culture and Mythology 89

Myths and legends surrounding volcanoes.. 89
Volcanoes and their symbolism 91
Volcanoes in art and literature.. 92

Tourism and the Exploitation of Volcanic Resources............... 94

Major Tourist Sites 94
Visiting a Volcano: Advice and Precautions....................................... 95
Ecotourism and Sustainable Volcano Management..................... 96
Geothermal and Mineral Resources 98

Conclusion.. 99

Future Challenges for Volcanology... 99
Acknowledgment 102

Introduction

Definition of a volcano

The definition of a volcano may seem obvious, but it is important to understand that volcanoes are much more than just mountains spewing fire. A volcano is an opening in the Earth's surface through which magma, gases, and volcanic ash can be ejected during an eruption.

Volcanoes are often found along the boundaries of tectonic plates, where the Earth's crust is most fragile. When two plates collide, one subducts beneath the other, creating a subduction zone where lava can accumulate and form a volcano.

There are different types of volcanoes, each with its own characteristics. Shield volcanoes, for example, are vast mountains with gentle slopes formed by the accumulation of fluid lava flows. Stratovolcanoes, on the other hand, are steep, conical mountains formed by the accumulation of ash and viscous lava.

The history of volcanoes dates back millions of years, when the Earth was still forming. Volcanoes have played a significant role in the geological history of our planet, shaping the landscape and contributing to the formation of the Earth's atmosphere.

In summary, a volcano is an opening in the Earth's surface through which magma, gases, and volcanic ash can be ejected. Volcanoes have played a significant role in Earth's history and are present in many locations around the world.

History of Volcanology

The history of volcanology goes back thousands of years, when the first volcanic eruptions were observed by humans. However, it was only in the past few centuries that volcanology emerged as a full-fledged scientific discipline.

The study of volcanoes began with the observations of naturalists and geologists such as Pliny the Elder and James Hutton, who proposed theories on the formation and origin of volcanoes. In the 18th century, scientists such as Lazzaro Spallanzani started to study eruptions in a more systematic way and began collecting samples of volcanic rocks for analysis.

In the 19th century, volcanology gained momentum with Georg von Neumayer's expedition to Antarctica in 1882, where he studied eruptions of Mount Erebus, and with the eruption of Mount Pelée in Martinique in 1902, which resulted in the death of over 30,000 people. These events prompted scientists to better understand volcanoes and develop methods to study them more precisely.

In the 20th century, technological advancements allowed volcanologists to study volcanoes more effectively, using tools such as seismographs, satellites, and drones. This has led to a better understanding of the processes that occur during eruptions, as well as the internal structure of volcanoes.

Today, volcanology is a constantly evolving discipline, with new discoveries and monitoring techniques being developed every year. Volcanologists work closely with governments and local communities to monitor active volcanoes, predict eruptions, and develop emergency plans in case of volcanic disasters.

Importance of Volcanoes in Earth's History

Volcanoes have played a crucial role in Earth's history, shaping our planet by creating unique landscapes, natural resources, and habitats for a variety of species. Since the formation of our planet, volcanoes have been major drivers in the evolution of Earth's crust and have contributed to the creation of new materials for the formation of igneous rocks, minerals, gases, and other elements.

Volcanoes have also created favorable conditions for life by providing nutrients for ecosystems and creating habitats for species that have evolved to survive in extreme environments. For example, the hot springs around volcanoes often serve as habitats for unique organisms that have evolved to survive in high temperature and high-pressure environments.

Furthermore, volcanoes have had a significant influence on the chemistry of the Earth's atmosphere by emitting gases such as carbon dioxide, sulfur, and hydrogen chloride, which have had effects on climate and the environment. Massive volcanic eruptions can have a substantial impact on the global climate by releasing massive amounts of particles into the atmosphere that can block sunlight and cool the planet.

However, volcanic eruptions can also warm the climate by emitting greenhouse gases, such as carbon dioxide.

The history of the Earth has been marked by significant volcanic events, such as the massive eruptions at Yellowstone around 640,000 years ago, which led to the formation of calderas and basalt plateaus. More recent volcanic eruptions have also left their mark on human history, such as the eruption of Mount St. Helens in 1980, which caused loss of life and considerable material damage.

Understanding volcanoes and how they operate is of paramount importance for human safety. Scientists can use knowledge of volcanoes to predict future eruptions and assist populations living in high-risk areas in preparing for catastrophic events. Indeed, a volcanic eruption can cause significant damage, such as the destruction of homes and infrastructure, loss of life, and economic and social disruption.

Distribution of Volcanoes Worldwide

The distribution of volcanoes worldwide is a fascinating and complex subject, influenced by various geological and geophysical factors. The presence of volcanoes is closely related to plate tectonics, which is the primary mechanism for the formation of Earth's landforms.

Volcanoes are typically found along the boundaries of tectonic plates, where compressional and divergent forces create weak zones in the Earth's crust. In these areas, the pressure accumulated beneath the Earth's surface can be released in the form of volcanic eruptions.

The Pacific Ring of Fire is one of the most active regions in terms of volcanic and seismic activity. This region stretches for approximately 40,000 kilometers around the Pacific Ocean and includes the west coast of South America, Japan, the Philippines, and Indonesia. Volcanoes in this region are typically stratovolcanoes and can be highly explosive, such as Mount Pinatubo in the Philippines or Mount St. Helens in the United States.

Outside of the Ring of Fire, there are also active volcanoes in Iceland, Ethiopia, and Indonesia. In Iceland, the high volcanic activity is due to the country's geological position, situated on the Mid-Atlantic Ridge, a zone of tectonic plate divergence. In Ethiopia, there is a large hotspot that generates regular volcanic activity, while in Indonesia, volcanoes are predominantly located on the islands of Java and Sumatra, which lie on a subduction zone.

Submarine volcanoes are also common and are primarily found along oceanic ridges, where the Earth's crust diverges, allowing for the emission of lava in the form of submarine volcanoes. These volcanoes have played a significant role in the formation of the ocean floor and the evolution of marine life.

Understanding the distribution of volcanoes worldwide is important for gaining a better understanding of their impact on the environment and society, as well as for better monitoring and prediction. Volcanoes can have dramatic consequences for populations living nearby, causing explosive eruptions, lahars, lava flows, and pyroclastic flows. Therefore, it is essential to develop effective monitoring and warning systems, as well as evacuation plans for affected populations.

Basics of Volcanology and Volcano Formation

Structure of the Earth

The Earth is a fascinating and complex planet, with an internal structure composed of different layers. Understanding this structure is essential for comprehending the volcanic phenomena that have shaped our planet for millions of years.

The Earth's crust is the outermost layer of the Earth and represents the surface on which we live. It is primarily composed of solid and rigid rocks, divided into two main types: continental crust and oceanic crust. The continental crust is thicker than the oceanic crust and consists mainly of granitic rocks, while the oceanic crust is predominantly composed of basaltic rocks. This difference in composition greatly influences the formation and activity of volcanoes.

The Earth's mantle is the layer located just beneath the crust and accounts for about 84% of the Earth's mass. The mantle is primarily composed of hot and viscous rocks that can deform under the effects of pressure and heat. This layer is divided into two parts: the upper mantle and the lower mantle. The boundary between these two parts is known as the Gutenberg Discontinuity.

The Earth's core is the deepest and central part of the Earth, representing approximately 15% of the planet's mass. It is mainly composed of iron and nickel and is divided into two parts: the liquid outer core and the solid inner core. This division is related to the convection of the liquid metal in the outer core, which generates a significant and vital magnetic field for life on Earth.

The structure of the Earth is fundamental to understanding volcanology, as volcanoes form due to tectonic and magmatic activity associated with the Earth's internal structure. Volcanoes are often associated with subduction zones, divergent boundaries, or hotspots, where the Earth's crust is thinner and magma can more easily rise to the surface. Therefore, knowledge of the Earth's structure is indispensable for understanding the origin and evolution of volcanoes.

Plate Tectonics

Plate tectonics is a central concept in volcanology, as it is responsible for the formation of volcanoes and volcanic eruptions. This theory was developed during the 20th century and is now widely accepted by the scientific community.

Plate tectonics describes the movement of lithospheric plates on the Earth's surface. The lithosphere is the rigid outer layer of the Earth, which includes the Earth's crust and the upper part of the mantle. Lithospheric plates are rigid blocks of rock that float on the asthenosphere, a softer layer located beneath the lithosphere.

Plate movement is caused by forces acting within the Earth,

such as mantle convection and gravitational forces. Plates can move relative to each other in three main ways: divergent, convergent, and transform boundaries.

Divergent boundaries are found where plates move away from each other. At these boundaries, magma rises to the Earth's surface and forms new oceanic crust or fissure volcanoes, such as those found along mid-ocean ridges.

Convergent boundaries are found where plates move towards each other. Convergent boundaries can be of two types: subduction or collision. In a subduction zone, an oceanic plate dives beneath a continental plate or another oceanic plate. Subduction can result in the formation of stratovolcanoes, such as those found in the Cascade Range in North America. In a collision zone, two continental plates collide and form mountains without generating volcanoes.

Transform boundaries are found where plates slide past each other, such as the San Andreas Fault in California. Transform boundaries generally do not generate volcanoes.

Plate tectonics is crucial for understanding the distribution of volcanoes on the Earth's surface. Volcanoes are typically located along divergent and convergent plate boundaries, as these are the areas where magma can rise to the surface.

Formation of Hotspots

The formation of hotspots is a complex and fascinating volcanic phenomenon that has long intrigued scientists. Hotspots are geothermal areas located in the Earth's mantle where heat flow is intense, creating conditions conducive to the formation of volcanoes. These volcanoes are often located beneath moving tectonic plates, resulting in a chain of volcanic islands. However, hotspots can also form underwater, creating underwater mountains.

Hotspots are considered geothermal anomalies, as they are located far from the usual boundaries of tectonic plates where volcanic activity typically occurs. Although their origin is still a subject of debate, a popular theory suggests that hotspots are remnants of the Earth's formation approximately 4.5 billion years ago. At that time, energy generated by asteroid and comet impacts melted the Earth, creating a vast ocean of molten magma. Hotspots would be remnants of that distant era.

Hotspots are aptly named because they occur in areas where the Earth's crust is thinner, allowing magma to more easily rise to the surface. This creates favorable conditions for the formation of volcanoes, which can erupt for years or even centuries. The Hawaiian Islands are a famous example of this phenomenon. The chain of volcanic islands was formed by the movement of the Pacific Plate over the hotspot located beneath the Island of Hawaii, leading to the formation of several volcanoes, including Kilauea and Mauna Loa, two of the most active volcanoes on the planet.

Scientists have been studying hotspots for decades and continue to discover new hotspots around the world. In addition to volcanic islands, hotspots can also cause underwater eruptions, creating submerged mountains that do not breach the water's surface. Scientists have found numerous underwater mountains, including the Pacific Ocean's submarine mountain range, which stretches over 43,000 kilometers.

The formation of hotspots is a fascinating example of how the Earth is in constant evolution. Scientists continue to study hotspots to better understand their origin and role in the formation of our planet. This research is of great importance for understanding the history of the Earth and for effectively predicting volcanic eruptions, which can have significant impacts on populations and the environment.

Magmatism and Volcano Formation

Magmatism is the process of magma formation and ascension beneath the Earth's surface, which leads to the formation of volcanoes. Volcanoes primarily form at the boundaries of tectonic plates, where the Earth's crust is thinner, and rocks are hotter. Plates can separate, converge, or slide past each other, causing movements of magma that give rise to volcanoes.

Magma is primarily composed of molten rocks but can also contain dissolved gases and crystals. Its composition varies depending on the region and depth at which it forms, as well as the type of volcano. Magma forms through the partial melting of rocks within the Earth's crust, either due to increased temperature, decreased pressure, or the addition of an external fluid like water.

Once magma is formed, it begins to rise toward the surface, propelled by the dissolved gases within the magma and the pressure exerted by the molten rock. When magma reaches the surface, it is called lava and begins to solidify, forming a volcano. The shape and size of the volcano depend on several factors, such as the viscosity of the lava, the amount of dissolved gases, and the speed of magma ascent.

Volcanoes can also form from hotspots, which are localized areas in the Earth's crust where magma is continuously present. Hotspots can be found beneath oceanic or continental crusts and can give rise to shield volcanoes, such as the volcanoes in Hawaii.

Igneous Rocks and Minerals

In volcanology, igneous rocks play a crucial role as they are closely associated with the formation of volcanoes. Igneous rocks are formed through the solidification of magma, the molten material inside the Earth that rises to the surface during a volcanic eruption.

Igneous rocks can be classified into two major categories: plutonic rocks and volcanic rocks. Plutonic rocks are formed through the slow solidification of magma within the Earth's crust, while volcanic rocks are formed through the rapid solidification of magma at the surface.

Plutonic rocks have larger grains and a more crystalline structure compared to volcanic rocks because they have had more time to form and cool slowly. Examples of plutonic rocks include granites, syenites, and gabbros.

Volcanic rocks, on the other hand, have finer grains and a more glassy or porphyritic structure because they cooled quickly at the surface. Examples of volcanic rocks include basalts, andesites, and rhyolites.

In addition to igneous rocks, there is also a wide variety of minerals found in volcanoes. Minerals are naturally occurring chemical compounds that form distinctive crystals with unique physical and chemical properties. Some common volcanic minerals include pyroxene, amphibole, plagioclase, olivine, and quartz.

The chemical composition of volcanic rocks and minerals is a key factor in determining the type of volcano and the resulting volcanic eruption. Shield volcanoes are typically composed of basalt, while stratovolcanoes are often composed of andesite or dacite. Rhyolites are associated with cinder cone volcanoes and calderas.

Composition of Volcanic Rocks

The composition of volcanic rocks is a fascinating subject in volcanology as it can vary significantly based on several factors, including geographical location, volcano type, and eruptive activity. Volcanic rocks are primarily composed of silicates, which are minerals containing silicon and oxygen, as well as other elements such as magnesium, iron, and calcium.

Volcanic rocks can be classified into two major categories: acidic volcanic rocks and basic volcanic rocks. Acidic volcanic rocks, such as rhyolites and dacites, have a high silica content and tend to be more viscous, which can result in explosive eruptions. Basic volcanic rocks, such as basalts, have a lower silica content and tend to be more fluid, which can result in effusive eruptions.

Silica content is one of the most important factors determining the type of volcanic rock formed and the associated eruptive activity. Magmas with high silica content tend to be more viscous, which can lead to the accumulation of pressure beneath the volcano, eventually resulting in explosive eruptions. Explosive eruptions are often the most hazardous to surrounding communities, as they can produce pyroclastic flows, ash clouds, and lahars that can cover vast areas.

Conversely, magmas with low silica content are more fluid and can produce effusive eruptions, such as lava flows. Effusive eruptions are generally less hazardous than explosive eruptions, as lava can be more easily predicted and controlled, but they can still cause considerable damage to surrounding communities.

Magmas producing volcanic rocks can also contain gases such as water, carbon dioxide, and sulfur dioxide. When magma reaches the surface, these gases can escape and form volcanic plumes and ash clouds. These gases can have negative effects on human health and the environment, particularly when emitted in large quantities.

It is also important to note that the composition of volcanic rocks can have significant implications for the environment. Volcanic ash can cause respiratory problems and impact air quality, while volcanic gases can have negative effects on health and the environment. Volcanic eruptions can also impact global climate, as volcanic aerosols can reflect sunlight, resulting in temporary cooling of the planet.

Magma and Lava: Differences and Properties

Understanding the difference between magma and lava is crucial to volcanology. Magma is molten rock located beneath the Earth's surface, while lava is magma that reaches the surface and solidifies. In other words, lava is magma that is ejected by a volcano during an eruption.

Magma is primarily composed of silicates, dissolved gases, and water. Its composition can vary depending on its depth and volcanic activity. Dissolved gases, such as carbon dioxide, sulfur dioxide, and hydrogen chloride, are responsible for volcanic explosions.

Lava is also primarily composed of silicates, but its composition can vary depending on the type of volcano and eruptive activity. The different properties of lava, such as viscosity, temperature, and chemical composition, can influence the effect of the eruption on the environment and surrounding populations.

The viscosity of lava, or its resistance to flow, is determined by the proportion of silicates it contains. The higher the silicate content, the more viscous the lava. Silicate-poor lavas tend to be more fluid and move more quickly, while silicate-rich lavas tend to be more pasty and move more slowly.

The temperature of lava can also vary significantly depending on the type of eruption. Effusive eruptions, characterized by low violence, typically produce hotter lavas, while explosive eruptions, characterized by high violence, generally produce cooler lavas.

The chemical composition of lava can also influence the impact of the eruption on the environment and surrounding populations. Gas-rich lavas, such as basaltic lavas, can produce pyroclastic flows, ash clouds, and volcanic explosions. Silicate-rich lavas, such as andesitic lavas, can produce lava flows and lahars.

Understanding the difference between magma and lava is essential for comprehending volcanic phenomena and predicting eruptions. Volcanologists study the composition, viscosity, and temperature of magma to predict future eruptions and assess associated risks. Scientists can also analyze the properties of lava to better understand the consequences of the eruption on the environment and surrounding populations.

Types of Volcanoes on Earth and Their Characteristics

Shield Volcanoes

Shield volcanoes, also known as Hawaiian volcanoes, are volcanoes that have a wide and flat shield-shaped form, with gentle slopes and rather calm eruptions. They are formed by effusive eruptions, where the fluid magma emerges smoothly from the magma chamber and flows down the volcano's flank. The eruptions are often long and continuous, with fluid lava flows that can travel great distances before solidifying.

Shield volcanoes are mainly found along oceanic rift zones, where tectonic plates separate and allow magma to rise to the surface. The most famous example is Mauna Loa, located in Hawaii, which is the world's largest shield volcano in terms of volume and altitude. Other notable shield volcanoes include Kilauea, also in Hawaii, and Piton de la Fournaise, on the island of La Reunion.

Shield volcanoes are important as they provide insights into submarine volcanic activity, as their structure is similar to that of undersea volcanoes. They are also significant for geology, as they can reveal information about the evolution of the Earth's mantle and magma composition. Additionally, they have significant economic implications as their lava can be used as raw material in road and building construction, and their tourism is a major source of income for local communities.

However, shield volcanoes are not without risks. Although their eruptions are generally less explosive than those of other types of volcanoes, lava flows can cause significant damage to surrounding infrastructure and homes. Moreover, emitted volcanic gases can be hazardous to human health, and eruptions can result in phenomena such as lahars and volcanic tsunamis.

Stratovolcanoes

Stratovolcanoes, also known as composite volcanoes, are among the most famous and dangerous in the world. These volcanoes have a complex structure composed of layers of ash, tephra, lava, and solid volcanic rocks. Eruptions from these volcanoes can be extremely explosive and produce devastating pyroclastic flows that can extend for kilometers.

These volcanoes form along subduction zones where one tectonic plate dives beneath another. During subduction, water is released from minerals in the descending plate, creating lighter magma that rises to the surface and forms volcanoes. Stratovolcanoes have a characteristic conical shape with a steep summit and a wider base.

Some of the world's most famous stratovolcanoes include Mount Fuji in Japan, Mount Saint Helens in the United States, and Mount Pinatubo in the Philippines. Mount Fuji, which reaches over 3,700 meters in altitude, is an important cultural symbol for the Japanese people for centuries. Mount Saint Helens experienced a catastrophic eruption in 1980, resulting in the deaths of 57 people and devastating nearby forested areas. Mount Pinatubo, on the other hand, had an explosive eruption in 1991 that had global climate repercussions due to the significant amount of ash and volcanic gases emitted into the atmosphere.

The eruptions of these volcanoes have had significant impacts on local populations and the environment. Lava flows and volcanic ash can disrupt local ecosystems and affect the quality of air and water. Volcanic gases emitted by eruptions can also affect the ozone layer and have consequences for the climate.

Monitoring stratovolcanoes is therefore essential for predicting eruptions and minimizing risks. Scientists use techniques such as seismic monitoring, measuring volcanic gas emissions, and satellite imaging to monitor volcanic activity. For example, monitoring Mount Saint Helens allowed for the prediction of its 1980 eruption, enabling authorities to implement safety measures and save many lives.

Lastly, stratovolcanoes are also important sites for scientific research. Volcanic rocks and sediments deposited during eruptions can provide insights into the Earth's geological history, including the formation of tectonic plates and the evolution of life. Stratovolcanoes can also be used to study the effects of volcanic eruptions on the environment and human health.

Scoria Cone Volcanoes

Scoria cone volcanoes, also known as strombolian volcanoes, are a common type of volcano found worldwide. As their name suggests, these volcanoes have a conical shape that primarily consists of cinders and scoria. Scoria cone volcanoes are created by explosive eruptions that eject significant amounts of ash, scoria, and rocks from the volcano's vent.

Eruptions from scoria cone volcanoes are typically of low intensity, producing relatively short and less fluid lava flows. However, these flows can still cause considerable damage and represent a danger to surrounding populations. Therefore, monitoring and predicting eruptions from these volcanoes are of great importance to ensure the safety of nearby communities.

These volcanoes can be found on the flanks of larger volcanoes, as secondary vents, or as individual volcanoes. They are often compared to barbecue chimneys, where lava is replaced by glowing embers that are ejected in all directions. Although less impressive than other types of volcanoes, scoria cones can still cause considerable damage and represent a danger to surrounding populations.

Scoria cone volcanoes have had a significant impact on human communities throughout history. The remains of the Roman city of Pompeii, which was destroyed by an eruption of Mount Vesuvius in 79 AD, are a famous example of how volcanic eruptions can change the course of history. More recently, the 1980 eruption of Mount St. Helens devastated a large part of Washington State, causing casualties and significant material losses.

Despite their potential danger, scoria cone volcanoes continue to attract tourists from around the world due to their beauty and accessibility. Volcanoes on the island of Hawaii, in particular, are famous for their spectacular beauty and accessibility. Visitors, however, must exercise caution and respect the warnings and restrictions imposed by local authorities.

Due to their explosive nature, scoria cone volcanoes are particularly unpredictable. Eruptions can be triggered by subtle changes in pressure or magma composition, making their monitoring challenging. Despite this, scientists have developed sophisticated tools to monitor volcanoes and predict eruptions. Volcanic gas sensors, for example, can detect changes in the composition of gases emitted by a volcano, providing clues about magma movements.

Submarine Volcanoes

Submarine volcanoes are often an unknown and fascinating aspect of our planet. They can be found in the oceans, often near plate divergence zones, where the oceanic crust is thinner and volcanic eruptions are more likely.

These volcanoes are formed similarly to those on land, but underwater conditions differ, resulting in unique geological formations that are often very different from terrestrial volcanoes. Submarine volcanoes are mainly fissure volcanoes, where hot lava escapes through faults in the oceanic crust and rapidly solidifies upon contact with cold water, forming basaltic columns and pillow lava. These geological formations are often stranger and more bizarre than terrestrial volcanoes, as they are influenced by the unique properties of water.

Submarine volcanoes have a significant impact on the marine environment. Submarine eruptions can release huge amounts of gas, ash, and lava into the water, creating large-scale changes in underwater ecosystems. Submarine lava flows can also alter ocean currents, seafloors, and create new habitats for organisms that can survive in these extreme conditions.

Submarine volcanoes also play a crucial role in the formation of islands and archipelagos in the oceans. For example, the Galapagos Islands and Hawaii were formed by submarine volcanoes that emerged from the water and rose above sea level. However, most submarine volcanoes are not large enough to emerge from the water and thus remain submerged.

Monitoring and predicting submarine eruptions are particularly challenging due to the inaccessibility of submarine volcanoes. Scientists use instruments such as pressure sensors, seismometers, and cameras to remotely monitor submarine volcanoes. However, these techniques are not always sufficient for accurately predicting submarine eruptions. Researchers are constantly searching for new methods to study submarine volcanoes and understand their behavior.

Finally, submarine volcanoes also have significant implications for the exploitation of marine resources. Submarine hydrothermal fields, which form around submarine volcanoes, can be sources of minerals and precious metals, but their exploitation can also have negative environmental consequences.

Rift Volcanoes and Fissure Eruptions

Rift volcanoes form in areas of tectonic plate divergence, where the Earth's crust stretches and fractures, creating faults. These faults allow magma to escape from the depths of the Earth and rise to the surface, thus forming volcanoes.

Rift volcanoes are primarily found in regions where tectonic plates diverge, such as the East African Rift, the North Atlantic Rift system, and the South Atlantic Rift system. These regions are characterized by frequent earthquakes and significant volcanic activity.

Eruptions from rift volcanoes are generally effusive, meaning that the magma is fluid and can flow easily over long distances. As a result, the eruptions are less explosive than those of other types of volcanoes, such as stratovolcanoes.

Lava flows from rift volcanoes are often spectacular, extending over kilometers and creating impressive volcanic landscapes. Rift volcanoes have also been associated with lava fountains that can reach impressive heights.

Fissure eruptions, on the other hand, are linear openings

in the Earth's crust from which an eruption occurs. These eruptions can occur in volcanic regions that are actively forming, such as in Iceland and Hawaii.

Eruptions from fissure eruptions are also often effusive, producing lava flows that can traverse long distances. These eruptions are less explosive than those of stratovolcanoes, but can still cause significant damage.

Rift volcanoes and fissure eruptions are important for volcanology as they provide a better understanding of plate tectonics and the evolution of the Earth's crust. Additionally, eruptions from these volcanoes have a significant impact on the environment, affecting the air and water quality as well as local ecosystems.

In summary, rift volcanoes and fissure eruptions are fascinating volcanic structures that offer many insights into the dynamics of the Earth and the geological processes at work in our planet, as well as the complex interactions between volcanoes, plate tectonics, and local ecosystems.

Mud Volcanoes

Mud volcanoes, also known as mudpots or fango volcanoes, are interesting and often overlooked geological phenomena. Unlike traditional volcanoes, which produce lava eruptions, mud volcanoes erupt with mud, water, and volcanic gases.

These volcanoes can be found worldwide, but most commonly in areas of active plate tectonics or geologically unstable regions. Mud volcanoes form when groundwater mixes with volcanic sediments and gases. Under pressure, this mixture can reach the surface, resulting in a mud eruption.

Mud eruptions can vary significantly in size and duration, ranging from small eruptions that last a few hours to larger eruptions that can last months or even years. The volcanic mud produced by these eruptions is typically composed of ash, sand, and organic matter. The emitted gases can include methane, hydrogen sulfide, nitrogen, and carbon dioxide, among others. These gases can be hazardous to human health and local ecosystems, as they can be toxic or suffocating.

Mud volcanoes have a significant impact on the local environment. They can alter the topography of the land, create new bodies of water, affect the quality of drinking water, and disrupt surrounding ecosystems. Mud eruptions can also cause damage to local infrastructure, such as roads, bridges, buildings, and water distribution networks.

Despite these risks, mud volcanoes often attract tourists due to their unique nature and accessibility. They can be safely visited if proper precautions are taken and safety guidelines are followed. Local authorities have implemented measures to protect visitors and local populations, including regular monitoring of mud volcanoes and the implementation of evacuation plans in case of emergencies.

Apart from their tourist appeal, mud volcanoes are also important for scientific research. Mud eruptions can help scientists better understand the geological processes involved in volcano formation and the nature of the Earth's crust.

Furthermore, mud volcanoes can also be used to harness geothermal energy. The steam and hot water reservoirs associated with these volcanoes can be used to generate electricity. This renewable and clean source of energy can contribute to reducing greenhouse gas emissions and support the transition to a low-carbon economy.

Extraterrestrial Volcanoes

Volcanism on the Moon

Volcanism is not exclusive to Earth, and the Moon is a fascinating example. The surface of the Moon is covered in craters, mountains, valleys, and other geological features that have been shaped by various processes, including volcanism. Although the Moon is no longer volcanically active, its volcanic history has played a significant role in its formation and evolution.

Volcanism on the Moon is the result of a combination of factors, including past geological activity, the chemical composition of the Moon, and the absence of an atmosphere. Volcanic eruptions on the Moon have been different from those on Earth due to these factors.

Volcanoes on the Moon are generally shield volcanoes, dome-shaped volcanoes, or cinder cone volcanoes. Shield volcanoes are the largest and most common, characterized by fluid lava eruptions that flow over great distances before cooling and hardening. Dome-shaped volcanoes are smaller and more compact, with viscous lava eruptions that cool and harden more quickly. Cinder cone volcanoes are smaller and more explosive, with ash and pyroclastic material eruptions.

The largest known volcano in the solar system, Olympus Mons, is on Mars, but the Moon also has impressive volcanoes. For example, Mount Piton on the visible face of the Moon measures approximately 2,200 meters in height and is surrounded by a 15-kilometer-wide caldera.

Volcanic eruptions on the Moon have produced volcanic materials such as lunar dust, rock, and minerals that have been studied and analyzed by lunar exploration missions. These materials have provided important information about the formation and evolution of the Moon, as well as volcanic processes in general.

Volcanism on the Moon has also impacted its climate and lunar environment. Volcanic gas emissions likely had an impact on the lunar atmosphere, although it is very thin. Lava flows have altered the lunar surface, creating unique landscapes and geological features.

Ultimately, volcanism on the Moon is a fascinating aspect of the geological history of our solar system. Although the Moon is no longer volcanically active, studying its volcanism can help us better understand volcanic processes on Earth and throughout the solar system.

Volcanism on Mars

Volcanism on Mars is a key feature of the red planet that has fascinated scientists and astronomy enthusiasts for decades. Volcanoes on Mars are very different from those on Earth in terms of size and composition, making them a fascinating subject of study for volcanologists and geologists.

The red planet is home to several famous volcanoes, including the giant shield volcanoes in the Tharsis region. The largest of them, Olympus Mons, is the largest known volcano in our solar system, reaching a height of 22 kilometers and having a base that covers an area larger than the entirety of Arizona. Tharsis volcanoes have a flat and wide dome shape, and their eruptions are primarily effusive, meaning that the lava flows slowly and peacefully down the volcano's slopes.

But not all volcanoes on Mars have the same shape. There are also stratovolcanoes on Mars that are similar to those found on Earth. These volcanoes are smaller than Tharsis volcanoes, but they are still impressive, reaching several kilometers in height.

Volcanoes on Mars are primarily made up of basalt, a common volcanic rock on Earth. However, the lava that flows from Mars' volcanoes is different from that on Earth. With Mars' lower gravity compared to Earth, lava has a harder time flowing, resulting in wider and less steep volcanoes.

The cause of volcanic activity on Mars is similar to that on Earth, which is due to plate tectonics. However, on Mars, this activity is related to a weakness zone on its crust called the «Tharsis rift zone». This rift zone allowed lava to flow easily and form large lava fields, which are common features on the Martian surface.

Volcanism on Mars also has interesting implications for the search for life on other planets. Volcanoes have the ability to create unique environmental conditions that could favor the emergence and survival of life. Additionally, the lava emitted by Martian volcanoes contains minerals and elements that could be useful for the production of construction materials and fuels on the planet.

Finally, studying volcanism on Mars can help understand the history and geology of the red planet, as well as the implications for the search for life and human colonization of Mars. Martian volcanoes provide a unique insight into volcanic activity in our solar system and offer an exciting opportunity for scientific research and space exploration.

Volcanism on Io

Io is a moon of Jupiter, considered the most volcanically active in the solar system. This moon was discovered by Galileo Galilei in 1610. It has a hot interior and powerful tidal forces exerted by Jupiter and neighboring moons that keep it in constant volcanic activity.

Volcanism on Io is mainly fueled by silicate magma erupting from cracks and fractures in the moon's crust. Eruptions can reach heights of several kilometers in the moon's tenuous atmosphere, producing lava flows, ash ejecta, and gas plumes that spread over hundreds of kilometers in space. Most of Io's volcanoes are located on its most active surface, near its equator.

Scientists have observed a variety of volcano types on Io, including shield volcanoes, calderas, cinder cones, and fissure volcanoes. Lava flows on Io are particularly remarkable due to their length, reaching hundreds of kilometers. Eruptions on Io are often associated with strange weather phenomena, such as the formation of sulfur dioxide clouds and other gases moving at high speeds around the moon. Eruptions can also generate lightning that illuminates the night sky.

Studying volcanism on Io can provide important information on how volcanoes function in other parts of the solar system. Active volcanoes on Io help us understand how geological processes can occur in extreme environments and how tidal forces can affect orbiting celestial bodies. Observations of Io's surface have also improved our understanding of the moon's chemistry and internal structure.

In addition to volcanoes, Io is also home to pools of sulfur and gas geysers. Sulfur pools are basins filled with molten sulfur, appearing as bright patches on the moon's surface. Gas geysers are jets of gas that rise up to 300 kilometers above the moon's surface.

Other Examples of Volcanism in Our Solar System

When we think of volcanoes, most of us immediately think of erupting mountains on Earth, but did you know that other celestial bodies also have volcanoes? The solar system is indeed filled with fascinating and unique volcanic phenomena.

Enceladus, one of Saturn's moons, also has active geysers of ice and water vapor. Scientists believe these geysers are fueled by liquid water reservoirs below the surface, which are heated by the moon's internal warmth.

Triton, Neptune's largest moon, also exhibits volcanic activity. Scientists believe eruptions on Triton are caused by internal heat generated by gravitational deformation of the moon.

Venus, often regarded as Earth's twin planet, is also covered in volcanoes. Scientists have discovered hundreds of active and inactive volcanoes on Venus' surface, which are larger and more numerous than those on Earth.

Titan, Saturn's largest moon, has also shown signs of volcanic activity. Scientists have found volcano-like structures on Titan's surface, but there is not yet definitive evidence of their activity.

Mimas, a small moon of Saturn, has a giant crater called Herschel that strangely resembles a cone-shaped volcano. However, scientists believe the crater was formed by the impact of an asteroid rather than volcanic activity.

Lastly, Ganymede, Jupiter's largest moon, has displayed signs of past volcanic activity. Scientists have discovered structures resembling lava flows on Ganymede's surface, which could indicate past volcanic activity.

These examples of volcanoes in our solar system serve as a reminder that volcanic phenomena are universal and not limited to Earth. They are the result of complex and fascinating geological and astronomical forces shaping our solar system today.

Volcanic Forms and Structures

Volcanic Cones

Volcanic cones are cone-shaped structures formed from explosive eruptions of viscous magma and solid rock fragments called tephra. These eruptions produce pyroclastic flows, ash, volcanic gases, and can extend for kilometers around. Volcanic cones can reach several kilometers in height and are often associated with violent and destructive eruptions.

Volcanic cones are typically built in multiple stages, with layers of tephra and lava gradually accumulating around the central crater. Eruptions can last for months or even years, and the volume of ejected material can be enormous. Volcanic cones are often located on convergent tectonic plates, where subduction of Earth's crust creates conditions favorable for magma formation.

Volcanic cones are fascinating natural phenomena, but they can also be very dangerous for populations living nearby. Eruptions can result in lahars, mudflows composed of ash and rocks that can devastate surrounding regions. Volcanic ash can also affect global climate by blocking solar radiation and cooling the Earth's surface.

Due to their destructive potential, volcanic cones are closely monitored by volcanologists. Measures such as seismic monitoring, volcanic gas measurement, and mapping of high-risk areas are used to prevent disasters and protect local populations. Despite the dangers associated with volcanic cones, they can also provide opportunities for ecotourism and scientific research.

Lava Domes

In this section, we will explore calderas, a unique and fascinating volcanic form. Calderas are large circular or arc-shaped depressions that form when the summit of a volcano collapses upon itself following a massive volcanic eruption. This process can occur rapidly or slowly over time, but in all cases, it leaves behind a spectacular caldera that can stretch for tens of kilometers.

Calderas are found in many volcanic regions worldwide, including Iceland, Hawaii, Indonesia, Japan, and the Andes. They can form from all types of volcanoes, including shield volcanoes, stratovolcanoes, and fissure volcanoes.

Calderas can vary significantly in size and shape, but they all have one thing in common: they offer a unique glimpse into the volcanic history of the region. The exposed layers of rocks and ash in the caldera walls can help volcanologists trace the eruptive history of the volcano and determine when and how it collapsed.

Calderas are also fascinating places to study geothermal phenomena. Many calderas are filled with lakes, hot springs, and fumaroles, which indicate the presence of magma and heat beneath the surface. These phenomena can be studied to better understand the geological processes at work beneath the Earth's surface.

In addition to their scientific interest, calderas are also popular destinations for tourists. Famous calderas like the Yellowstone Caldera in the United States attract millions of visitors each year. It is important to note that calderas are hazardous environments, and visitors must adhere to strict safety guidelines to avoid risks of eruption, landslides, and toxic volcanic gases.

Lava Flows

Lava flows are one of the most fascinating phenomena of volcanic eruptions. A lava flow is a stream of magma that flows on the ground, usually at temperatures exceeding 1000°C. Lava flows can take different forms depending on the viscosity of the magma and the topography of the eruptive area.

Lava flows can be divided into two main types: effusive flows and explosive flows. Effusive flows are the most common and occur when the magma is fluid and flows easily. These flows are typically slow but can cover great distances and last for weeks or even months. Explosive flows, on the other hand, occur when the magma is more viscous and builds up pressure inside the volcano. When this pressure is released, it can cause an explosion that propels fragments of magma and rock in all directions.

Lava flows have a major impact on the environment. They can destroy entire cities and villages, block rivers, create new landscapes, and even modify the climate in the long term. Lava flows can also have beneficial effects by enriching soils and creating new habitats for plants and animals.

Lava flows can be fascinating to watch, but it is important to remember that they can be extremely dangerous. It is essential to follow the instructions of local authorities in the event of a volcanic eruption and never approach a lava flow without proper training and equipment.

Basalt Plateaus

Basalt plateaus are extensive volcanic structures often associated with hotspots. These vast areas of basaltic rock are formed by the accumulation of numerous lava flows emitted from volcanic fissures over long periods of time.

One of the most well-known examples of a basalt plateau is the Deccan region in India. This plateau covers an area of nearly 500,000 square kilometers and is the result of intense volcanic activity that occurred approximately 65 million years ago. This eruption had a major impact on terrestrial biodiversity, likely contributing to the extinction of dinosaurs.

Another important example is the Columbia Plateau in the northwestern United States. This plateau is composed of layers of basalt that were emitted approximately 17 million years ago during the activity of the Yellowstone hotspot. The Columbia Plateau is now a fertile region for agriculture and livestock.

Basalt plateaus are also present on other planets, including Mars, where areas such as Tharsis and Elysium are examples of extensive volcanic regions. Studying these plateaus on other planets can provide important information about plate tectonics and the geological evolution of celestial bodies.

Basalt plateaus can be valuable resources for humanity, particularly due to the presence of rare metals and minerals in basaltic lava. However, the exploitation of these resources can have significant environmental consequences, including air and water pollution.

Finally, the formation of basalt plateaus is an important subject of research in volcanology. Understanding the processes that lead to the formation of these volcanic structures can help predict future eruptions and better understand the geological evolution of our planet.

Evolution of a Volcano over Time

The evolution of a volcano over time is a complex process that can span thousands or even millions of years. It all begins with the formation of a magma chamber beneath the Earth's surface. This magma chamber is filled with magma, a mixture of molten rocks, gases, and crystals.

Over time, the pressure in the magma chamber increases, which can cause a volcanic eruption. When the magma reaches the Earth's surface, it cools and solidifies, forming volcanic rocks such as lava, ash, and scoria.

After an eruption, the volcano may enter a period of rest, where it can remain inactive for years or even centuries. During this time, the magma chamber may refill again with magma, preparing for the next eruption.

Over time, volcanic eruptions can change the structure of the volcano. Lava flows can form new layers of volcanic rocks, which can cover the old ones. Explosive eruptions can create craters and calderas, which can collapse and create depressions on the volcano's flanks.

Time can also affect the composition of volcanic rocks. Metamorphism and metasomatism processes can alter the composition of rocks, transforming existing minerals and incorporating new chemical elements into the rock.

The evolution of a volcano is also influenced by its geographical location. Volcanoes located on hotspots can be more stable and have a more uniform shape, while volcanoes located on subduction zones can be more unstable and have a more irregular shape.

Lastly, the evolution of a volcano is also influenced by human activities. Mining, dam construction, and geothermal activities can disrupt the structure of a volcano and affect its eruptive behavior.

Volcanic Products

Types of Lava

Volcanoes produce a wide variety of lavas with unique characteristics that depend on the conditions of their formation and eruption. It is important to understand the different types of lava in order to better grasp the characteristics of volcanic eruptions and the potential hazards they pose.

Basaltic lavas are the most common and are produced by shield volcanoes. They have low viscosity, allowing them to flow easily over long distances and form basaltic plateaus. Basaltic lava flows are often accompanied by lava fountains that erupt from the volcano's summit.

Andesitic lavas are more viscous than basaltic lavas and are often produced by stratovolcanoes. They can form lava domes that accumulate on the volcano's surface. These domes can be very unstable and cause explosive eruptions.

Rhyolitic lavas are the most viscous and explosive. They are often produced by calderas and supervolcanoes. Rhyolitic eruptions produce viscous lava flows, pyroclastic explosions, and pyroclastic flows that can devastate large areas.

In addition to the three main types of lavas, there are also phonolitic and trachytic lavas that fall between basaltic and rhyolitic lavas in terms of viscosity and chemical composition. These lavas are often produced by volcanoes with unique chemical compositions that are not common.

Understanding the different types of lava is crucial for predicting volcanic eruptions and assessing the risks associated with each type of eruption. Basaltic lavas can cause fires and damage to infrastructure, while andesitic and rhyolitic lavas can be more destructive and disrupt regions or even the entire world.

Lava Tubes

Lava tubes are geological wonders that form when molten lava flows beneath a solid crust on the surface, creating underground passages that can extend for several kilometers. These tubes are often associated with shield volcanoes, as these have effusive eruptions that produce fluid lava flows capable of traveling long distances.

First explored in the 1700s in Iceland, lava tubes have now become popular tourist attractions in many volcanic regions around the world. In addition to being tourist sites, lava tubes offer incredible opportunities for scientists to study volcanic processes and the Earth's geology.

Lava tubes are the result of complex volcanic processes that can be thoroughly studied through these unique structures. Visitors have the chance to discover the beautiful colors and shapes of these underground formations, while scientists can study the geological characteristics of the tubes to better understand the eruptive processes that created the volcanoes.

In addition to their scientific value, lava tubes can also provide opportunities for ecotourism and biodiversity preservation. Studies have revealed the presence of unique microorganisms and fauna specifically adapted to lava tubes, making them fascinating ecosystems to explore.

However, visiting lava tubes can be dangerous due to the presence of toxic gases and risks of collapse. Visitors should always be accompanied by experienced professional guides to minimize risks and learn to appreciate these wonders safely.

Lava tubes also have significant implications for geothermal energy, as they can serve as conduits for heat flow beneath the Earth's surface. Lava tubes can be used to harness geothermal heat and produce renewable energy, offering sustainable and environmentally friendly alternatives to meet society's energy needs.

Lastly, lava tubes have a significant impact on the cultures and myths of many countries. In many cultures, volcanoes are considered sacred and mystical places. Lava tubes are often associated with spiritual beliefs and myths that date back thousands of years.

Tephra and Volcanic Ash

In the field of volcanology, tephra and volcanic ash are important products of volcanic eruptions. Tephra refers to rock fragments, ranging in size from sand grains to boulders, that are ejected during a volcanic eruption. Volcanic ash is very fine tephra, the size of dust particles. Tephra and volcanic ash can be projected to considerable altitudes and transported over long distances by winds.

These volcanic products can have significant effects on local populations, ecosystems, and infrastructure. Tephra can cause damage to buildings and crops, as well as endanger human and animal life. Volcanic ash can affect respiratory systems and eyes, disrupt transportation systems, communications, and water supplies.

Tephra and volcanic ash can also have consequences for the environment and climate. Volcanic ash particles in the atmosphere can block sunlight and cool the climate for several years. Volcanic tephra can alter the morphology of landscapes and create new geological formations.

Predicting volcanic eruptions and associated tephra and volcanic ash is a major challenge for volcanologists. Tephra and volcanic ash are often projected to high altitudes, making their detection and monitoring difficult. Numerical models can help predict the behavior of tephra and volcanic ash, but these models are based on simplifying assumptions and may not always provide precise forecasts.

In terms of volcanic risk management, mapping of hazard zones is crucial for preventing the negative effects of tephra and volcanic ash on local populations. Emergency and evacuation plans must be developed to ensure the safety of populations and minimize material losses. Critical infrastructure, such as airports, must also be equipped to deal with tephra and volcanic ash.

Volcanic Gases

Volcanoes are fascinating phenomena that have a significant impact on our environment and climate. Volcanic eruptions release large amounts of gases, which can have dramatic consequences for populations and ecosystems near volcanoes. Volcanic gases are complex mixtures of different components, including water vapor, carbon dioxide, sulfur, and halogens.

The composition and quantity of volcanic gases vary depending on magma characteristics, eruption conditions, and local geology. Explosive volcanoes, such as Mount St. Helens, produce gases rich in sulfur dioxide and fine particles, which can have long-term effects on air quality and human health. Effusive volcanoes, like Kilauea, mainly emit water vapor and carbon dioxide, which have more limited effects on the environment.

Volcanic gases can have direct impacts on populations near volcanoes. Volcanic eruptions can release massive amounts of toxic gases like hydrochloric acid and hydrogen fluoride, which can cause chemical burns, eye injuries, and respiratory problems. Eruptions can also result in pyroclastic flows, lahars, and volcanic tsunamis, which can be extremely dangerous for populations located in hazard zones.

Volcanic gases can also have broader impacts. Emissions of sulfur dioxide and fine particles can cause acid rain, negatively affecting air and surface water quality. Emissions of volcanic gases can also contribute to climate cooling by reflecting some of the sunlight and reducing the amount of light reaching the Earth's surface.

Lahars and Mudflows

Lahars and mudflows are devastating phenomena that can occur during volcanic eruptions and pose considerable risks to populations and infrastructure near volcanoes.

Lahars are volcanic mudflows that form when rainfall, snowmelt, or melted ice mixes with ash, debris, and eruptive material and cascades down the slopes of volcanoes at high speeds. Lahars can be extremely fast, reaching speeds of several tens of kilometers per hour, and can carry rocks, trees, and other heavy materials that can cause considerable damage.

Mudflows are flowing streams of liquid mud that form when a large amount of rainfall or snowmelt infiltrates eruptive deposits and triggers a landslide. Mudflows can be extremely hazardous as they can move rapidly and sweep away anything in their path.

Lahars and mudflows are common phenomena during volcanic eruptions, and it is essential to monitor and predict them to reduce risks for populations and infrastructure near volcanoes. Monitoring techniques include the use of radar to measure surface deformations of the volcano, observation of debris cones to detect signs of ground movement, and analysis of seismic data to detect signs of volcanic activity.

Forecasting lahars and mudflows is a complex challenge,

as these phenomena can occur rapidly and unpredictably. Computer models can be used to simulate the behavior of lahars and mudflows and to predict their path, but their reliability depends on the quality of input data and the accuracy of parameters.

Risk management for lahars and mudflows involves mapping hazard zones, urban planning to avoid critical infrastructure construction in exposed areas, and the establishment of alert systems and evacuation plans to protect populations in case of emergency.

Pyroclastic Flows and Ejections

Pyroclastic flows and ejections are volcanic phenomena that inspire both fear and awe due to their destructive power and spectacular visuals. These phenomena are among the most devastating observed during a volcanic eruption, capable of causing considerable damage to local populations, infrastructure, habitats, and agricultural land.

A pyroclastic flow, also known as a glowing avalanche, is a mix of hot gases and rock fragments that can race down a volcano's slopes at incredible speeds, sometimes exceeding 700 km/h. Ejections, on the other hand, are solid materials, such as ash and fragments of lava, that are expelled into the air during an eruption and can fall onto surrounding areas.

Pyroclastic flows form when a lava dome or a growing volcano collapses, releasing hot gases and rock fragments. When the lava dome becomes too unstable, it collapses onto itself, creating an avalanche of volcanic debris that moves at considerable speeds. Pyroclastic flows can travel distances of up to 50 km and overcome obstacles, such as hills and mountains, to reach lower and more populated areas.

Ejections, on the other hand, are expelled into the air when a volcano erupts, propelled by the pressure of gases within the magma. Ejections can vary in size, ranging from the size of a grain of sand to that of a house, and their speed and trajectory depend on many factors, such as magma composition and eruption strength. Ejections can be projected to considerable heights, reaching several kilometers, and be transported over long distances by winds, threatening areas hundreds of kilometers away from the erupting volcano.

To prevent the dramatic consequences of these phenomena, it is important to understand their formation mechanisms and behaviors. Scientists have developed techniques to monitor these phenomena, such as using sensors to measure pressure and temperature of gases emitted by volcanoes, as well as modeling techniques to predict the trajectory and dispersion of pyroclastic flows and ejections. This monitoring allows for better eruption forecasts and the evacuation of at-risk populations.

Despite these scientific advancements, pyroclastic flows and ejections remain unpredictable and challenging to anticipate, and it is therefore essential to take precautionary measures to protect populations living in volcanic hazard zones.

Processes of Volcanic Eruptions

Effusive Eruptions

Effusive eruptions are a type of volcanic eruption where lava flows relatively slowly, without producing violent explosions. These eruptions can last for weeks or even years and are characterized by the formation of lava flows that can extend over long distances.

Effusive volcanoes are mainly found along rift zones and hotspots, where low-viscosity magma rises to the surface. Effusive eruptions occur when the magma is mainly composed of basalt, a volcanic rock with a high silica content.

When the magma reaches the surface, it can form lava flows that can extend for kilometers, slowly but relentlessly. Lava flows can be divided into two types: pahoehoe flows, smooth and shiny, and aa flows, rough and jagged.

Effusive eruptions can be dangerous for local populations, especially if lava flows threaten cities or villages. Lava flows can cause considerable damage to infrastructure, agriculture, and inhabited areas. Volcanic gases can also be hazardous to health, particularly for those with respiratory problems.

However, effusive eruptions can also be beneficial for the environment. Lava flows can enrich the soil with minerals, promoting plant and tree growth. Moreover, effusive eruptions can help prevent explosive eruptions by releasing magma pressure in the volcano.

Explosive Eruptions

Explosive eruptions are among the most violent and destructive volcanic events. They are characterized by the emission of large amounts of pyroclastic materials, such as ash, pumice, blocks, and volcanic bombs, as well as highly pressurized volcanic gases. Explosive eruptions can cause significant damage to property, infrastructure, and loss of human life.

The violence of explosive eruptions is mainly due to the presence of viscous and gas-rich magma. When the gas pressure reaches a critical level, magma is explosively ejected, creating ash and gas columns that can reach several kilometers in height. These columns can extend for hundreds of kilometers and affect the inhabited areas below them.

Explosive eruptions can occur in different types of volcanoes, including stratovolcanoes and shield volcanoes. Stratovolcanoes, also known as cone volcanoes, are particularly prone to explosive eruptions due to their complex structure and viscous magma. Shield volcanoes, on the other hand, tend to have more effusive eruptions as their magma is more fluid and has low gas content.

Explosive eruptions can also lead to phenomena such as pyroclastic flows, nuee ardentes, lahars, and volcanic tsunamis. Pyroclastic flows are swift streams of hot gas and pyroclastic materials that descend rapidly down the volcano's slopes at speeds exceeding 100 km/h.

Nuee ardentes are pyroclastic flows that travel longer

distances and can reach inhabited areas below the volcano. Lahars are volcanic mudflows that form when a mixture of water, ash, and volcanic blocks race down the volcano's slopes, causing considerable damage in downstream valleys. Volcanic tsunamis are waves triggered by volcanic eruptions that can cause considerable damage to nearby coasts.

Fumaroles and Solfatara

Fumaroles and solfatara are spectacular and fascinating volcanic manifestations. They are often present in active volcanoes and indicate the presence of hot and acidic gases escaping from the ground. Fumaroles are emissions of steam, while solfatara are emissions of sulfuric gases.

These phenomena are often associated with volcanic activity and can be observed on the flanks and craters of volcanoes. Fumaroles and solfatara are formed by hot gases produced by magmatic processes inside the volcano. These gases propagate through rocks and rise to the surface, where they are released into the atmosphere.

Fumaroles and solfatara can be very dangerous to living beings due to the presence of toxic gases such as sulfur dioxide and hydrochloric acid. Scientists study these phenomena to better understand volcanic activity and monitor volcanic eruptions.

Fumaroles and solfatara can also have consequences on the environment. The emitted gases can contribute to the acidification of soils and water, negatively affecting fauna and flora. Volcanic eruptions can also have detrimental effects on local populations and regional economies.

Despite the dangers they pose, fumaroles and solfatara are fascinating volcanic manifestations that testify to the power and complexity of geological phenomena. As custodians of the Earth, volcanologists are committed to the study and understanding of these phenomena to better predict volcanic eruptions and protect populations and ecosystems.

Volcanic Tsunamis

Volcanic tsunamis are devastating phenomena that can be triggered by subsea or coastal volcanic eruptions. They are caused by a combination of factors, including volcanic explosions, flank collapses, and submarine landslides. These events can generate giant waves that propagate over hundreds of kilometers, posing a threat to coastal communities and infrastructure.

The deadliest tsunami in history, the 2004 Indian Ocean tsunami, was triggered by an underwater earthquake. However, volcanic tsunamis can also be highly destructive. In 1883, the eruption of Krakatoa in Indonesia generated a tsunami that killed over 36,000 people.

Volcanic tsunamis can also have significant ecological impacts. The giant waves can transport volcanic materials, such as rocks, ash, and sediments, into coastal areas. These materials can affect marine habitats and coastal ecosystems, including mangroves and coral reefs.

It is important to monitor active volcanoes and coastal areas for precursor signs of possible volcanic tsunamis. Warning systems and evacuation plans must be developed to help coastal communities prepare for such events. Volcanology research should continue to explore the mechanisms governing volcanic tsunamis for a better understanding of their behavior and predictability.

Factors Influencing Eruption Types

Volcanic eruptions are complex phenomena that can be influenced by many factors. Several elements can contribute to determining the nature and severity of the eruption, including magma composition, internal volcano pressure, and the presence of dissolved gases in the magma.

The type of magma present in the volcano can play a crucial role in the type of eruption that will occur. Magma contains dissolved gases that escape as bubbles when pressure decreases. If the magma is viscous, the bubbles struggle to escape, which can lead to increased pressure and explosive eruptions. Conversely, if the magma is fluid, gases can easily escape and cause effusive eruptions.

Internal volcano pressure can also play a significant role in the type of eruption that will occur. High pressure can result in violent explosions, while lower pressure allows magma to flow more easily out of the volcano.

Dissolved gases in magma can also play an important role in the type of eruption that will occur. Gases such as water, carbon dioxide, and sulfur can cause explosions if trapped in magma and unable to escape. If magma is gas-rich, it can lead to more violent and explosive eruptions.

Other factors can also influence the type of eruption that will occur, such as volcano topography, seismic activity in the region, and the composition of the surrounding rock.

Ultimately, the complexity of volcanic eruptions means that many factors can influence the type of eruption that will occur. Understanding these factors is crucial for predicting eruptions and reducing risks for populations living near active volcanoes.

Famous Volcanoes and Historical Eruptions

Mount Vesuvius and Pompeii

Mount Vesuvius is one of the most famous volcanoes in the world due to its tumultuous history and proximity to the ancient city of Pompeii. Located in the Campania region of Italy, it is considered one of the most dangerous active volcanoes. Mount Vesuvius stands at a height of 1,281 meters and its last major eruption occurred in 1944.

The eruptive history of Mount Vesuvius began approximately 25,000 years ago, with explosive eruptions producing pyroclastic flows and tephra. Since then, the volcano has experienced numerous eruptions, but the one in 79 A.D. was the most famous and devastating.

This eruption was so violent that it destroyed the cities of Pompeii, Herculaneum, and Stabiae within a matter of hours, burying their inhabitants under layers of ash and stones. Approximately 16,000 people were killed during this eruption.

Since then, Mount Vesuvius has had several eruptions, with the most recent one occurring in 1944. This eruption resulted in the deaths of 26 people and caused damage to the surrounding towns.

Mount Vesuvius is closely monitored by scientists and surveillance teams to prevent future major eruptions. The area around the volcano is also closely monitored, and local authorities have developed evacuation plans for residents in case of an eruption.

Due to its history and proximity to the city of Pompeii, Mount Vesuvius is a major tourist site in Italy. Visitors can admire the beauty of the volcano and learn more about its eruptive history. However, it is important to respect safety rules and follow the advice of local authorities in case of danger.

Mount Pelée and Saint-Pierre

Mount Pelée is an iconic volcano located in the Caribbean on the island of Martinique. This mountain, towering over 1,300 meters, is one of the most active volcanoes in the Lesser Antilles arc and was the stage for a major eruption in 1902 that profoundly marked the history of volcanology.

This eruption began on April 23, 1902, with precursor signs such as gas and ash emissions. Local authorities, concerned about the signs of volcanic activity, organized the evacuation of certain high-risk areas, but unfortunately underestimated the severity of the situation and did not evacuate the city of Saint-Pierre, which was located in close proximity to the volcano.

On May 8, 1902, the situation rapidly escalated. A cataclysmic eruption took place, ejecting pyroclastic clouds at hundreds of kilometers per hour that obliterated the city of Saint-Pierre and killed nearly 30,000 people within minutes. This catastrophe was one of the deadliest in modern volcanic history and raised global awareness of the need to better understand volcanoes and the associated risks.

The eruption of Mount Pelée had profound and lasting consequences on modern volcanology. It allowed scientists to better understand volcanic eruptive phenomena and develop techniques for monitoring and predicting eruptions. It also prompted the establishment of evacuation and protection plans for populations living near volcanoes, as well as the development of strategies to minimize the consequences of eruptions.

Today, Mount Pelée is closely monitored by scientists and local authorities to minimize risks for surrounding populations. Modern tools such as seismic stations, gas and ground deformation measurements, thermal cameras, etc., are used to detect precursor signs of a potential eruption and enable a swift alert to endangered populations.

The eruption of Mount Pelée has also had a considerable impact on the cultural and social history of Martinique and the Caribbean. It has inspired artists, poets, writers, and left an indelible mark in the collective memory of the region.

Krakatoa and its Global Effects

Krakatoa, located in the Sunda Strait between the islands of Java and Sumatra in Indonesia, is one of the most famous volcanoes in the world. This mountain entered history in 1883 when it experienced one of the most devastating eruptions ever recorded. This eruption had global effects and demonstrated how volcanoes can impact the environment and society.

The eruption of Krakatoa began on August 26, 1883, with violent explosions that emitted clouds of ash, rock, and gas at an altitude of over 20 kilometers. Tsunamis were triggered by the fall of ash into the water, causing gigantic waves that swept the coasts of Java and Sumatra. The explosions were heard over 4,800 kilometers away and projected ash into the atmosphere, resulting in glowing sunsets worldwide for several years.

The eruption was responsible for the deaths of at least 36,000 people, primarily due to the following tsunamis. The waves devastated coastlines, destroying entire villages and causing thousands of casualties. The ash also led to the death of numerous animals, causing significant damage to local ecosystems.

But the eruption of Krakatoa also had a significant impact on the global climate. The particles and gases emitted by the eruption cooled the planet for several years, leading to harsher winters and cooler summers in many regions of the world. Harvests were affected, and famines followed in some areas.

The eruption of Krakatoa marked a turning point in the history of volcanology by showing that volcanic eruptions can have global-scale effects. Since then, scientists have worked to better understand the mechanisms underlying volcanic eruptions and to develop monitoring and prediction methods to help minimize risks to local populations.

Today, Krakatoa is still considered an active volcano and is closely monitored. Scientists study its movements and behavior to detect any precursory signs of an imminent eruption. In the event of an eruption, local authorities are prepared to quickly evacuate local populations to minimize human losses.

Mount St. Helens and Modern Awareness

Mount St. Helens is a volcano located in the state of Washington, United States, which erupted on May 18, 1980. The eruption was one of the largest in U.S. history, causing the deaths of 57 people and millions of dollars in damages. This event sparked great interest in volcanology and volcanic risk management, leading to significant advances in these fields.

Modern awareness of volcanic risks has been greatly influenced by the eruption of Mount St. Helens. After the disaster, scientists worked hard to better understand volcanoes and ways to prevent associated catastrophes. The findings of this research have been used to improve volcanic risk prediction and management.

As a result, volcanic monitoring has been strengthened, and new alert systems have been put in place to inform the public in case of imminent danger. Public awareness has also improved, with awareness and education campaigns launched to inform people about risks and how to protect themselves.

The eruption of Mount St. Helens also led to a better understanding of volcanic eruptions. Scientists were able to observe the different phases of the eruption and the effects it had on the environment. These observations allowed for a better understanding of the processes that occur during a volcanic eruption and contributed to new advancements in the field of volcanology.

Finally, the eruption of Mount St. Helens had a significant impact on volcanic risk management worldwide. The lessons learned from this disaster have been applied to other active volcanoes around the world, helping to better predict and manage the risks associated with volcanic eruptions.

Other Famous Volcanoes

There are many famous volcanoes in the world that have left their mark on the planet's history. Some have caused devastating natural disasters, while others have become popular tourist destinations for their natural beauty.

Mount Fuji, located in Japan, is one of the most famous and iconic volcanoes in the world. It is considered one of the country's three most important symbols and attracts millions of visitors each year. Mount Fuji is a stratovolcano, reaching over 3,700 meters in height. It is considered an active volcano but has not had a major eruption since the early 18th century.

Mount Rainier, located in the state of Washington, United States, is another famous volcano due to its impressive size and natural beauty. It is an active stratovolcano, standing at over 4,300 meters in altitude. Although it has not had a major eruption in over a century, scientists closely monitor Mount Rainier due to its proximity to the city of Seattle.

Eyjafjallajökull, located in Iceland, gained some notoriety in 2010 due to its eruption, which caused the cancellation of thousands of flights in Europe because of the ash cloud it produced. Although the eruption did not cause major damage, it highlighted the potential impact of volcanic eruptions on travel and infrastructure.

Mount Kilimanjaro, located in Tanzania, is an inactive stratovolcano famous for its impressive height of over 5,800 meters. It is also known for its unique biodiversity and is a popular destination for hikers and tourists.

Mount Nyiragongo, located in the Democratic Republic of Congo, is an active stratovolcano famous for its ever-changing lava lake. This unique characteristic attracts scientists and visitors from around the world.

Lastly, Mount Etna, located in Sicily, is one of the most active volcanoes in the world, with a major eruption occurring on average every two to three years. This volcano has cultural and historical significance for the people of Sicily, who consider it a nature deity.

These famous volcanoes serve as a reminder of the power and beauty of nature, but they can also cause devastating natural disasters. That is why volcanic risk monitoring and management are so important in protecting populations living near these geological phenomena.

Monitoring and Forecasting Volcanic Eruptions

Tools and Techniques for Volcanic Monitoring

Volcanic monitoring is a crucial aspect of modern volcanology, allowing for the prediction of volcanic eruptions and the protection of nearby populations. There are various tools and techniques used to monitor volcanoes, each with its advantages and limitations.

One of the most commonly used monitoring tools is a network of seismometers, which record earth movements caused by magma movements beneath the surface. The data collected by these instruments help detect changes in seismic activity, which can indicate an imminent eruption. Seismic monitoring systems have been developed to be accessible online and in real-time, allowing volcanologists to remotely monitor volcanoes.

Another important tool is satellite imagery, which allows for the monitoring of volcanic activity from space. Satellite images can detect changes in temperature and other modifications within the volcano's environment, enabling volcanologists to monitor eruptions in real-time and observe the volcano's structural developments.

Measuring ground deformation is also a valuable tool for volcanic monitoring. Deformation measuring devices, such as inclinometers and extensometers, detect ground movements associated with volcanic activity. These instruments can be used to detect changes in shape, size, or volume of a volcano, as well as to forecast eruptions.

Monitoring the composition of volcanic gases is another technique used to detect imminent eruptions. Volcanic gases often contain high levels of carbon dioxide and sulfur, as well as other compounds such as hydrogen chloride, hydrochloric acid, and water. Changes in the composition of volcanic gases can indicate an upcoming eruption.

Finally, direct visual monitoring is also an important means of monitoring volcanic activity. Volcanologists can observe ash emissions, lava fountains, and explosions to track volcanic activity and forecast eruptions. Drones and aerial surveys are also visual monitoring means that enable precise and detailed data collection on volcanic activity.

Precursor Signs

Precursor signs of a volcanic eruption can vary significantly depending on the type of volcano and its level of activity. However, there are several common indicators that can help predict an imminent eruption.

First, changes in the shape or surface of the volcano can indicate volcanic activity. For instance, sudden swelling or deformation of the volcano may indicate an accumulation of magma beneath the surface. Fissures and fractures may also form in surrounding areas, releasing volcanic gases and steam.

Seismic activity can be a key indicator of volcanic activity. Frequent, low-intensity, shallow earthquakes can indicate magma movements beneath the surface. More intense earthquakes can indicate rock fractures, magma injection, or an impending eruption.

Furthermore, monitoring volcanic gases can aid in predicting an eruption. Volcanoes regularly emit volcanic gases, including carbon dioxide, sulfur, and sulfur dioxide. Sudden increases in the levels of these gases can indicate heightened volcanic activity.

Lastly, visual observation is also important. Fumaroles and hot springs can indicate volcanic activity. Changes in the color and composition of groundwater can also serve as indicators.

It is important to note that volcanic eruption monitoring and prediction is not an exact science, and precursor signs are not always clear. Often, a combination of multiple indicators is necessary to establish a model of volcanic behavior and predict an imminent eruption.

Eruption Modeling and Forecasting Methods

Modeling and forecasting volcanic eruptions are constantly evolving and continuously developing fields. Scientists employ a variety of techniques to study volcanoes and predict eruptions.

One of the primary methods used for eruption prediction is continuous monitoring of volcanoes using geophysical and geochemical sensors. These sensors can measure variations in pressure, temperature, and chemical composition of magma and volcanic gases. Field observations, such as the appearance of fractures and ground deformations, can also be used to detect early signs of an eruption.

Numerical models are another important tool for volcanic eruption prediction. Scientists can use mathematical models to simulate conditions within a volcano and predict the consequences of a potential eruption. Models can be used to forecast the path of lava flows, the dispersion of volcanic ash, and the effects of pyroclastic flows.

Eruption forecasting methods also include the use of remote sensing techniques, such as satellite imaging and lidar, for long-distance volcano monitoring. Remote sensing data can provide information about variations in the volcano's surface, magma temperature, and volcanic gas composition.

It is important to note that volcanic eruption prediction carries uncertainties and limitations. Scientists must take into account numerous variables that can affect the nature and severity of the eruption, including magma composition, source depth, water presence, and surrounding topography. Furthermore, even with sophisticated techniques, accurately predicting the timing, duration, and severity of an eruption can be challenging.

Despite these challenges, improving volcanic eruption prediction is essential for public safety and risk management. The information gathered by scientists can be used to inform crisis management decisions and help communities prepare for potential eruptions.

Limitations and Challenges in Prediction

Volcanic eruption prediction is a complex and challenging field. Although volcanology has made significant progress in monitoring and predicting eruptions, there is still much to be done to improve prediction accuracy and reduce risks for populations living near volcanoes.

One of the main limitations in volcanic eruption prediction is the uncertainty associated with volcanic processes themselves. Volcanoes are complex natural systems that can evolve rapidly and unpredictably. Even with the most advanced monitoring techniques, accurately predicting when and how an eruption will occur can be difficult.

Another limitation in volcanic eruption prediction is related to monitoring and data collection capabilities. While technology has greatly enhanced volcano monitoring, there are still regions of the world where monitoring is limited or nonexistent. This can make forecasting volcanic eruptions and managing risks for local populations challenging.

Furthermore, volcanic eruption prediction can be complicated by human factors, such as negligence or a lack of awareness among populations living near volcanoes. In many cases, people are reluctant to evacuate or follow safety guidelines, even when warned of an imminent eruption. This can lead to significant loss of life and property damage.

Despite these limitations, it is important to continue improving volcanic eruption prediction. This can be done by further developing new monitoring and data collection technologies, strengthening public awareness of volcanic hazards, improving collaboration between scientists and local authorities, and investing in research on volcanic processes.

Ultimately, volcanic eruption prediction is vital for protecting local populations, reducing material damages, and maintaining public safety. By working together to enhance our prediction capabilities, we can help prevent volcano-related disasters and gain a better understanding of the impact these natural phenomena have on our planet and society.

Volcanic Risk Management

Mapping of Volcanic Hazard Zones

Mapping volcanic hazard zones is a crucial element in volcanic risk management. It allows for the identification of areas that are potentially affected by volcanic eruptions, the planning of emergency interventions, and the minimization of loss of human lives and material assets. The mapping of hazard zones is based on the analysis of the geological, geomorphological, and volcanological characteristics of volcanoes and their surroundings.

The first step in mapping hazard zones is to establish detailed geological and volcanological databases for each volcano. These databases contain information on past eruption characteristics, current activity, eruption frequency, eruption size, eruption type, and potential hazards associated with each eruption type. These databases are then used to develop volcanic risk maps for each volcano.

Volcanic risk maps are created using advanced numerical modeling techniques that take into account the different characteristics of the volcano and its surroundings, such as topography, population density, vulnerability of infrastructure, and evacuation routes. The numerical models are often validated with field data and satellite observations.

The volcanic risk maps are used to identify areas that are potentially affected by different types of volcanic eruptions, including areas exposed to lava flows, pyroclastic flows, lahars, and ash fall. The risk maps are also used to establish emergency plans and evacuation plans to minimize loss of human lives and material assets in the event of an eruption.

Mapping volcanic hazard zones is a key element in volcanic risk management, but it is not without limitations. The numerical models used to create risk maps are based on simplifying assumptions that can introduce errors. Furthermore, volcanic eruptions are often unpredictable, making it challenging to plan for emergency interventions.

Urban Planning

Urban planning is a critical issue when it comes to managing volcanic risks. Many cities around the world are located near active volcanoes, exposing their populations to potential hazards. To minimize risks and protect populations, appropriate urban planning is needed.

First and foremost, it is important to map the hazard zones around volcanoes. This will help delimitate areas where construction is prohibited or where strict construction standards must be applied. Regulations should include measures for the construction of buildings that are resistant to earthquakes, lava flows, and volcanic ash.

It is also crucial to develop evacuation plans in case of an imminent eruption. These plans should be regularly tested and updated to ensure their effectiveness. Populations should be educated about volcanic hazards and the need to stay informed about volcanic activity. Alert systems should be put in place to quickly inform populations in case of imminent danger.

Urban planning should also take into account the economic and social consequences of a volcanic eruption. Economic losses can be substantial for affected areas. Therefore, contingency plans should be in place to assist affected populations and businesses in quickly recovering from the eruption.

Lastly, it is essential to implement sustainable development policies to avoid exacerbating volcanic risks. Human activities, such as mining, dam construction, or deforestation, can have detrimental effects on volcanic activity and increase risks for surrounding populations.

Education and Awareness

Raising public awareness and providing education about volcanoes are crucial aspects in ensuring the safety of populations living near these natural wonders. Understanding the risks associated with volcanoes is essential for making informed decisions and preparing for emergency situations.

One of the most effective methods for raising public awareness about volcanoes is to use simple and understandable analogies and metaphors for non-specialists. For example, one can use the analogy of a volcano as a boiling pot of water. Like the pot of water, a volcano can unpredictably explode, projecting steam, gases, lava, and hot debris in all directions. Understanding this analogy can help raise public awareness about the potential hazards of volcanoes.

It is also essential to explain to people how volcanoes are monitored and how scientists predict eruptions. Volcanic monitoring tools and techniques, such as deformation measurements, analysis of volcanic gases, and seismic observations, can help predict eruptions. It is important to emphasize that even with the most advanced technologies, volcanic eruptions remain difficult to predict with accuracy.

Furthermore, it is important to focus on volcanic risk management, which includes mapping hazard zones, urban planning, education and awareness, alert systems, evacuation plans, and crisis management. It is crucial to prepare populations living near volcanoes to face emergency situations and provide them with the necessary knowledge to make informed decisions in the event of an eruption.

Education about volcanoes should not be limited to the associated risks but should also include an understanding of the impact of volcanoes on the environment, climate, and life. Volcanoes have shaped the Earth's landscape and have had a significant impact on the evolution of life on our planet. Understanding these phenomena can help raise public awareness about the importance of preserving volcanoes and the ecosystems surrounding them.

Lastly, it is important to emphasize that education about volcanoes should not be restricted to populations living near volcanoes. Everyone can benefit from understanding these fascinating natural phenomena, and it can contribute to strengthening our connection with nature and developing our overall environmental awareness.

Alert Systems and Evacuation Plans

Surveillance and prediction of volcanic eruptions are key elements in volcanic risk management. Alert systems are used to warn populations in hazard zones of an imminent eruption. Evacuation plans are put in place to help people seek shelter and minimize loss of human lives.

Volcanic alert systems vary depending on countries and regions, but they usually include seismic monitoring, volcanic gas monitoring, visual surveillance, and ground deformation monitoring. Data collected from these techniques are analyzed in real-time, and alerts are issued when signs of an imminent eruption are detected.

Evacuation plans are developed in collaboration with local and regional authorities, emergency services, and local communities. They are designed to evacuate people from hazard zones before the eruption occurs, using predetermined evacuation routes and emergency shelter centers.

Evacuation plans must be clear, precise, and easily understandable for the general public. They should include information about evacuation routes, emergency shelters, actions to take in case of emergency, and priority for the evacuation of people and domestic animals.

However, it is important to highlight that predicting volcanic eruptions remains challenging and uncertain. Precursor signs can be ambiguous or unpredictable, and determining the exact timing and intensity of a volcanic eruption is often difficult. Evacuation plans can also be complicated by factors such as topography, weather conditions, and limited road infrastructure.

Despite these challenges, alert systems and evacuation plans remain key elements in volcanic risk management. They can help save lives and minimize material losses during a volcanic eruption. By raising awareness among local communities about volcanic hazards and developing robust emergency plans, we can help protect vulnerable populations from volcanic eruptions.

Crisis Management

Managing volcanic crises is a complex task involving multiple actors, from the local community to government organizations, scientific teams, and emergency services. Volcanic crises can take various forms, ranging from minor seismic activity to major eruptions that can cause significant damage.

The first step in volcanic crisis management is continuous volcano monitoring and assessment of potential risks to local populations. Scientists monitor volcanoes using a variety of techniques, such as seismic monitoring, volcanic gas monitoring, and ground deformation monitoring. Data collected through these techniques are analyzed to assess the level of threat that the volcano poses to the local population.

Once a volcano is considered dangerous, a series of measures are implemented to protect the population. Local and national authorities work together to develop emergency plans that outline the actions to be taken in the event of a volcanic eruption. These plans often include evacuation zones for people living near the volcano, as well as plans for the distribution of food, water, and other essential supplies in case of emergency.

Scientists also work closely with emergency services to help predict the consequences of volcanic eruptions. Computer models can be used to predict the direction of lava flows, ash fall, and pyroclastic flows, as well as estimate the amount of material ejected by a given eruption.

Managing volcanic crises is a challenging task, as volcanic eruptions can be unpredictable and devastating. Authorities must be prepared to act quickly in emergency situations while working on long-term efforts to minimize risks for local populations. Efforts to manage volcanic crises must be supported by international cooperation and open and transparent communication with local populations.

The Impact of Volcanoes on the Environment and Climate

Geological and Geomorphological Impacts

Volcanoes have a significant impact on Earth's geology and geomorphology. Volcanic eruptions can cause dramatic changes in the landscape, creating new relief features such as valleys, gorges, canyons, mountains, and plains. Volcanoes can also form islands and archipelagos in the oceans, altering the world map.

During eruptions, large quantities of materials like ashes, pumice stones, lapilli, and lava blocks can be ejected and deposited on surrounding soil, modifying its chemical and mineralogical composition. These materials can also affect rivers and local ecosystems, sometimes causing damage to infrastructure and homes.

Volcanoes also influence Earth's water cycle. When lava flows into the ocean, it rapidly cools and solidifies, creating new underwater rock formations. Additionally, volcanoes can produce hot springs, geysers, and fumaroles that influence local and regional weather conditions by generating clouds and precipitation.

Moreover, volcanic eruptions can have global environmental consequences. Volcanic ash and emitted gases can spread through the atmosphere, absorbing sunlight and temporarily cooling Earth's surface. This cooling can have significant effects on global climate, impacting plant growth, agricultural production, and animal migration patterns.

Volcanoes also play a crucial role in soil formation. Volcanic materials, such as ash and lava, are rich in minerals and essential nutrients for plant growth. These materials can be carried by rivers and winds, fertilizing agricultural and forested lands and improving crop yields and natural habitats.

The Impact of Volcanic Gas Emissions

Volcanic gas emissions can have significant impacts on the environment and climate. Volcanic gases include water vapor, carbon dioxide, nitrogen, sulfur, chlorine, and halogens, which can all affect air quality and regional and global climate conditions.

Carbon dioxide is the most abundant volcanic gas and is considered a major greenhouse gas. Volcanic carbon dioxide emissions can contribute to climate change by increasing the amount of greenhouse gases in the atmosphere. However, volcanic carbon dioxide emissions are relatively low compared to anthropogenic emissions from fossil fuel use.

Sulfur gases emitted by volcanoes can also have a significant environmental impact. Sulfur dioxide reacts with water and oxygen in the atmosphere to form sulfuric acid, which can cause acid rain. Acid rain can acidify lakes and rivers, damage soils and plants, and pose health risks to humans and animals.

Volcanic eruptions can also release ash particles and aerosols into the atmosphere. These particles can reflect sunlight, resulting in temporary cooling of Earth's surface. The eruption of Mount Pinatubo in 1991 caused a decrease in global average temperature of 0.5 degrees Celsius for several years.

In addition to environmental impacts, volcanic gas emissions can also affect human health. Volcanic gases can cause eye and respiratory irritations, headaches, and nausea in exposed individuals. Populations living near active volcanoes are particularly vulnerable to volcanic gas emissions.

Volcanic Aerosols and Climate Cooling

When a volcano erupts, it not only releases lava and volcanic gases but also aerosols, tiny particles that can remain suspended in the atmosphere for months or even years. These volcanic aerosols can have surprising effects on climate, ranging from local to regional or even global cooling.

Volcanic aerosols mainly consist of sulfur dioxide, which is converted to sulfuric acid once released into the atmosphere. These particles can partially or completely block sunlight, leading to a decrease in Earth's surface temperature. This has occurred several times in the past, notably after the eruption of Mount Pinatubo in the Philippines in 1991, which resulted in a global average temperature decrease of 0.5°C for about two years.

Volcanic aerosols can also affect cloud formation and

precipitation. Volcanic aerosols can act as condensation nuclei for water droplets, increasing cloud formation. This can lead to more abundant precipitation, but also more acidic precipitation due to the presence of sulfuric acid in the aerosols.

The impact of volcanic aerosols on climate depends on various factors, including the quantity and composition of the aerosols, their altitude in the atmosphere, the season, and the geographical location of the eruption. Additionally, volcanic aerosols can interact with other climate factors, such as greenhouse gases, ocean currents, and atmospheric winds, making their effect on climate hard to predict.

Volcanoes and the Carbon Cycle

Volcanoes play a crucial role in Earth's carbon cycle. Volcanic gas emissions, including carbon dioxide (CO_2), methane (CH_4), and nitrous oxide (N_2O), contribute to the increasing concentration of greenhouse gases in the atmosphere. However, volcanoes can also store carbon in volcanic rocks and marine sediments.

When a volcano erupts, it releases stored carbon dioxide from magma and surrounding rocks. The CO_2 rises into the atmosphere and can remain suspended for several years. Volcanic eruptions can thus have an impact on climate by temporarily increasing the concentration of greenhouse gases.

However, volcanoes can also store carbon. When magma

cools and solidifies, it forms mineral-rich volcanic rocks containing carbon, such as calcium carbonate. Volcanic rocks can then transform into marine sediments, where carbon is stored for millions of years.

Furthermore, volcanoes have a role in the carbon cycle through their influence on geological processes. Volcanic eruptions can cause changes in geochemical cycles, such as erosion, soil formation, and the production of essential nutrients for plant growth.

Finally, vegetation around volcanoes also plays a crucial role in the carbon cycle. Volcanic soils are often nutrient-rich and promote the growth of plants that absorb CO_2 from the atmosphere. Forests around volcanoes can, therefore, help regulate the carbon cycle by storing CO_2 in biomass.

Volcanoes and Life

Ecosystems around volcanoes

Ecosystems around volcanoes provide a unique environment that presents both challenges and opportunities for life. Volcanoes can be extremely inhospitable habitats due to their intense heat, toxicity, and instability, but they can also harbor a diversity of species adapted to these extreme conditions.

Ecosystems around volcanoes are often characterized by species that have developed unique adaptations to survive in harsh conditions. Some species have developed mechanisms to withstand high temperatures, toxic gases, nutrient-poor soils, and drought conditions. Plants around volcanoes can act as pioneers, rapidly colonizing new rocky surfaces, contributing to soil formation and vegetation expansion.

Ecosystems around volcanoes can also harbor special fauna. Insects, reptiles, and birds may be attracted to the warm, dry conditions of volcanic slopes. Some bat species have even been observed roosting in volcanic calderas. Large herbivores such as deer, elk, and goats may also be attracted to the grassy slopes of volcanoes, but their impact on the environment can be negative.

Active volcanoes can also offer opportunities for scientific research and the discovery of new species. Scientists can study volcano ecosystems to understand how life adapts to extreme conditions and find solutions to environmental challenges. Volcanic eruptions can also expose new mineral deposits, such as precious metals and rare minerals, which can be exploited for industrial applications.

However, the proximity of volcanoes can also pose risks to life. Volcanic eruptions can destroy habitats and ecosystems, resulting in the loss of many species. Lava flows and lahars can also damage surrounding lands, reducing the region's capacity to support life.

Organisms adapted to volcanic environments

Volcanoes are extremely hostile environments, marked by extreme temperatures, pressure, and chemical composition. Despite these challenges, some organisms have managed to adapt to these volcanic conditions to find refuge and survive. These organisms, known as extremophiles, are capable of surviving in extreme and often inhospitable environments, such as hydrothermal vents or acidic environments.

Organisms adapted to volcanic environments include a wide variety of microorganisms, such as bacteria and archaea, as well as plants, animals, and even humans. Volcanic microorganisms are often the first to colonize new volcanic surfaces, such as recent lava flows, volcanic ash, or fumaroles, and are often responsible for the formation of volcanic soils.

Plants are also able to adapt to these extreme environments, such as mosses, lichens, and ferns. These plants are often able to survive in nutrient-poor soils, rich in volcanic minerals such as iron, sulfur, and magnesium, as well as in environments with high levels of toxic volcanic gases such as sulfur dioxide.

Animals adapted to volcanic environments are also numerous and varied, including insects, spiders, crustaceans, fish, and birds. Birds are particularly remarkable for their ability to nest on steep volcanic cliffs, as is the case for terns and puffins. Fish and crustaceans have adapted to extreme hydrothermal vent environments, such as black smokers, which are often characterized by high temperatures and high levels of sulfur and heavy metals.

Finally, humans have also developed strategies to adapt to volcanic environments, including exploiting geothermal resources such as hot springs and geothermal fields for energy production or thermal activities. Humans have also learned to live with volcanic risks, developing monitoring and alert techniques, as well as evacuation plans to protect populations living near active volcanoes.

The role of volcanoes in species diversification

Volcanoes have had a significant impact on species diversification worldwide. Volcanic eruptions can disrupt existing habitats, but they can also create new habitats and provide opportunities for evolution.

When a volcano erupts, it can destroy surrounding habitats by covering the land with lava and ash. However, after an eruption, newly exposed areas develop, offering opportunities for colonization by species that could not establish themselves previously. Nutrient-rich volcanic ash can also stimulate plant growth and provide food for herbivores.

Volcanoes can also play an important role in the formation of new species. When groups of organisms are geographically isolated by a volcanic eruption, they can evolve separately and develop different characteristics. This process is called allopatric speciation and can be observed on isolated volcanic islands such as the Galapagos.

Volcanic environments can also select specific traits that favor species survival. Organisms adapted to high temperatures, acidic pH, and high levels of sulfur have a better chance of surviving in extreme volcanic environments.

Furthermore, volcanoes can also have indirect effects on species diversification by influencing the global climate. Large volcanic eruptions can release large amounts of gases and ash into the atmosphere, causing temporary global cooling. This cooling can impact diversification by causing sudden environmental changes and pushing species to adapt.

Volcanoes and the origin of life

Volcanoes have played a crucial role in the origin of life on Earth. Although this may seem paradoxical, as volcanic eruptions may appear hostile to life, they have actually provided favorable conditions for the emergence of life.

Firstly, volcanoes contributed to the formation of the Earth's primitive atmosphere, which was very different from the current atmosphere. Volcanic eruptions released large amounts of gases, such as water, carbon dioxide, ammonia, methane, and hydrogen sulfide, creating a primitive atmosphere on Earth. These gases formed clouds, which caused acid rain that helped erode rocks, thus forming the primitive oceans.

Furthermore, volcanoes also released essential minerals for life, such as iron, sulfur, magnesium, and phosphorus, which served as catalysts for the necessary chemical reactions for life. These minerals were also incorporated into living organisms, contributing to their growth and development.

Volcanoes also provided heat, which allowed chemical reactions to occur more rapidly, increasing the chances of organic molecule formation. Volcanic eruptions also released organic materials, such as amino acids, which are the building blocks of proteins. These organic materials were able to combine to form more complex molecules, such as nucleic acids, which are the building blocks of DNA.

Additionally, volcanoes created extreme environments, such as hot springs and geothermal areas, which provided ideal conditions for the emergence of life. These environments provided chemical energy in the form of temperature and concentration gradients, fueling the early forms of life. Bacteria and archaea were the first life forms to appear on Earth, and some of these life forms are still present today in extreme volcanic environments, such as hot springs and geothermal fields.

Scientists have discovered bacteria and archaea living in

extreme volcanic environments, which have metabolisms dependent on the chemical elements released by volcanoes. These life forms use heat and volcanic gases to produce chemical energy, which is used to power their metabolism. This discovery shows that volcanoes continue to play an important role in the evolution of life on Earth.

Volcanoes in Culture and Mythology

Myths and legends surrounding volcanoes

The myths and legends surrounding volcanoes have a fascinating history that dates back thousands of years and spans many cultures from around the world. Volcanoes have often been regarded as creatures or divine forces that could bring both life and death.

In Greek mythology, the god Hephaestus was considered the master of volcanoes and forging, responsible for volcanic activity. In Hawaiian mythology, the goddess Pele was revered as the goddess of fire and volcanoes, and the Hawaiians believed that eruptions were the result of her anger and emotions.

The legends and beliefs surrounding volcanoes have also inspired many stories and folk tales. For example, the Aztecs believed that volcanoes were gates to the underworld, where the souls of the dead were sent to be judged. In Polynesia, myths say that the volcano Mauna Kea on the island of Hawaii is the home of the god Poli'ahu, who is considered the guardian of winter and snow. The myths and legends surrounding volcanoes have also inspired artists throughout the centuries.

The German painter Caspar David Friedrich created numerous paintings depicting dramatic volcanic landscapes, while the

American photographer Ansel Adams captured incredible images of volcanic eruptions.

However, the myths and legends surrounding volcanoes have also had negative consequences. The ancient Romans often sacrificed human victims to volcano gods in order to appease their anger, while the people of Easter Island destroyed their own environment by building monumental statues around volcanoes.

Today, the myths and legends surrounding volcanoes continue to fascinate and inspire people. Volcanoes have also been a source of inspiration for filmmakers and documentary makers, who have created many fascinating films and documentaries on the subject.

Nevertheless, it is important to remember that science and research continue to provide new knowledge about volcanoes, and that this knowledge is essential for better understanding and managing these forces of nature. Volcanic eruptions can be extremely dangerous to nearby populations, as well as the environment. Scientific knowledge is therefore necessary to prevent and minimize the impacts of volcanic eruptions.

In essence, the myths and legends surrounding volcanoes bear witness to the fascination and awe that these natural phenomena have evoked in the minds of humans throughout history.

Volcanoes and their symbolism

Volcanoes have fascinated humanity for millennia. They have inspired many cultures around the world and have been used as symbols in art, literature, and mythology. In this section, we will explore the various symbolisms that volcanoes have taken on over the centuries.

In some cultures, volcanoes have been associated with the anger and vengeance of the gods. Violent and destructive eruptions were often interpreted as punishments for transgressions committed by humans. For example, the volcano Stromboli in Italy is considered the «Mouth of Hell» due to its frequent and spectacular eruptions that have terrified local populations.

In other cultures, volcanoes have been revered as sources of life and fertility. The ashes and minerals ejected by volcanoes were considered essential elements for the growth of plants and the fertility of the land. Volcanoes were also associated with creative and benevolent deities, such as the goddess Pele in Hawaiian mythology.

Volcanoes have also been used as symbols of transformation and renewal. Volcanic eruptions can destroy entire landscapes, but they can also create new terrains and unique forms of life. In Japanese culture, cherry blossoms that bloom on the slopes of Mount Fuji symbolize ephemeral beauty and regeneration after a catastrophe.

Finally, volcanoes have been associated with adventure and exploration. Many travelers have been drawn to the wild and majestic landscapes of volcanoes, creating a flourishing tourism industry. Hikers, mountaineers, and extreme sports enthusiasts are particularly attracted to active volcanoes, seeking to challenge their limits while discovering new horizons.

Volcanoes in art and literature

Volcanoes have always been a source of inspiration for artists and writers. They have fascinated humanity for centuries and have been used to represent symbolic and metaphorical ideas.

In art, volcanoes have been depicted in paintings, sculptures, engravings, and photographs. Artists have often sought to capture the dramatic and spectacular effect of volcanic eruptions, using vibrant colors and expressive forms to create powerful images.

For example, the artist William Hodges created magnificent paintings of erupting volcanoes during his travels through the South Pacific in the late 18th century. He managed to capture the immense power and striking beauty of volcanic eruptions.

In literature, volcanoes have often been used as symbols of destruction and chaos. In Jules Verne's famous novel «Journey to the Center of the Earth», a volcano is the entry point into the mysterious and dangerous world located beneath the Earth's surface. The volcano represents an imminent and inevitable threat here.

Volcanoes have also been used as metaphors to describe intense emotions or social conflicts. For example, in Victor Hugo's novel «Les Misérables», the 1832 revolution in Paris is compared to a volcanic eruption that disrupts and destroys everything in its path.

Lastly, volcanoes have also inspired poets, who have used their beauty and power to express strong emotions. In his famous poem «Kubla Khan», Samuel Taylor Coleridge describes an idyllic landscape with a volcano at its center, symbolizing creativity and imagination.

Tourism and the Exploitation of Volcanic Resources

Major Tourist Sites

Exploring major volcanic tourist sites is an incredible experience that attracts millions of people worldwide every year. These sites offer breathtaking views, unique landscapes, fascinating history, and insight into the power of nature.

One of the most famous sites is Yellowstone National Park, located primarily in the US state of Wyoming. This park offers stunning views of one of the world's largest geyser systems, including Old Faithful. Visitors can also explore colorful hot springs, bubbling mud pots, and volcanic landscapes shaped by lava eruptions and ash flows.

Mount Fuji in Japan is also a popular tourist site. It is the highest peak in Japan and an active volcano. Visitors can climb Mount Fuji during the summer season to enjoy breathtaking views of the surroundings. Moreover, the surrounding region is known for its natural hot springs called onsens, which provide visitors a way to relax after a day of hiking.

In Italy, Mount Etna is one of the most popular attractions on the island of Sicily. It is considered one of the most active volcanoes in the world, with frequent eruptions that have shaped the surrounding landscape. Visitors can climb the volcano to admire the view of the Sicilian coast or explore lava caves created by past eruptions.

Hawaii, a US state located in the Pacific, is also famous for its volcanoes. Hawaii Volcanoes National Park offers spectacular views of lava flows, craters, and volcanic fissures. Visitors can also explore lava caves, witness ongoing lava eruptions, and discover Hawaiian traditions and culture related to volcanoes.

In Iceland, Vatnajökull National Park is home to Europe's largest glacier and several active volcanoes. Visitors can explore ice caves, walk on glaciers, and admire spectacular waterfalls formed by past volcanic eruptions.

These sites are just a few examples among many other volcanic tourist sites worldwide. They offer unique experiences for curious travelers passionate about discovering the wonders of nature. However, it is important to remember that these sites can also pose dangers to visitors, and it is crucial to follow safety instructions and warnings issued by local authorities.

Visiting a Volcano: Advice and Precautions

When visiting a volcano, it is important to take certain precautions to ensure one's safety as well as that of others. Firstly, it is essential to gather information about the specific volcano one intends to visit. Volcanoes differ in their characteristics and associated risks, so knowing what precautions to take is important.

Next, it is recommended to visit the volcano with an experienced guide and strictly follow their instructions. The guide has good knowledge of the area and knows how to react in case of emergencies. They can also provide interesting information about the volcano, its history, and geological features.

It is also important to equip oneself with appropriate safety gear such as sturdy hiking boots, weather-appropriate clothing, and a helmet. In some cases, a respiratory mask may be necessary to protect against volcanic gas emissions.

It is further recommended not to travel alone and to stay in groups with fellow visitors. In case of any problem, it is easier to seek help when accompanied.

Finally, it is important to respect the rules established by local authorities, such as prohibited areas or access restrictions. Volcanoes can be dangerous and unpredictable, so following these rules is crucial for ensuring one's safety.

Ecotourism and Sustainable Volcano Management

Volcanoes attract millions of visitors each year, but the management of tourism in these areas can cause considerable environmental damage. This is why it is important to adopt an ecological and sustainable approach to preserve these unique and fragile sites.

The first step is to assess the impact of tourism on the

environment. Visitors can disturb natural habitats, damage landscapes by leaving waste, trampling plants, and disturbing wildlife. Therefore, it is essential to limit the number of visitors, define routes, and provide resting areas to minimize disturbances.

Furthermore, it is important to raise visitor awareness about precautions to minimize their impact on the environment. Information should be provided on the behaviors to adopt, such as not leaving any waste, respecting prohibited areas, and staying on marked trails. Experienced guides can help visitors better understand the environment and local culture while minimizing negative impacts.

Waste management is also a crucial element of sustainable volcano management. Waste should be collected and responsibly disposed of to avoid soil, river, and water pollution. Organic waste can be composted, while recyclable waste can be collected for processing elsewhere. Recycling and waste management programs can also be implemented to raise visitor awareness and encourage environmentally friendly behaviors.

Finally, sustainable volcano management may also involve the use of renewable energy sources to minimize environmental impact. For example, geothermal energy can be used to generate electricity or heat buildings, reducing dependence on non-renewable energy sources and decreasing greenhouse gas emissions.

Geothermal and Mineral Resources

Volcanoes are not only natural wonders but also offer important resources for humanity. Geothermal and mineral resources are two examples of the benefits provided by volcanoes.

Geothermal energy is the use of heat from the Earth to generate electricity and heat. Volcanoes are a significant source of geothermal energy, as their activity produces a considerable amount of heat that can be harnessed and utilized. This heat is often used to power geothermal plants that generate electricity and heat for homes and industrial buildings.

Additionally, volcanoes are also sources of minerals such as gold, silver, copper, zinc, and many more. These minerals are often found in magmatic and hydrothermal rocks formed by volcanic activity. Mines extracting these minerals are often located in active or ancient volcanic areas.

Volcanoes can also serve as sources of construction materials such as pumice stone, lava, and basalt. These materials are often used for building roads, retaining walls, and buildings in areas near volcanoes.

However, the exploitation of geothermal and mineral resources from volcanoes must be carried out with caution to avoid environmental damage and risks to human health. Mining activities can cause environmental damage, and geothermal energy production may also result in greenhouse gas emissions.

Conclusion

Future Challenges for Volcanology

Volcanology is a constantly evolving science, with new challenges to overcome in order to understand and predict volcanic eruptions, as well as to protect populations and ecosystems near volcanoes. Here are some of the future challenges for volcanology:

Improved prediction of volcanic eruptions:

Predicting volcanic eruptions accurately is a complex task that requires in-depth knowledge of volcano characteristics, precursor signals, and eruptive processes. Scientists must continue to develop new methods to detect early signs of an eruption, such as variations in seismic pressure, ground deformation, or gases emitted by the volcano. Prediction models also need to be enhanced to better forecast volcanic eruption behaviors, taking into account variations in magma composition, terrain topography, and external factors like weather conditions.

Enhanced volcano monitoring:

Volcano monitoring is crucial for preventing volcanic eruptions and protecting local populations. Scientists must continue to develop new tools and technologies to monitor volcanoes, like drones, satellites, and sensor networks, and improve interoperability among different monitoring platforms. Furthermore, training and preparedness of volcano monitoring teams need to be strengthened to ensure their effectiveness in emergency situations.

Understanding the impact of volcanic eruptions on climate:

Volcanic eruptions have a significant impact on the global climate due to the amount of gases and particles they release into the atmosphere. Scientists must continue studying the complex interactions between volcanic eruptions and the climate system, using computer models to simulate the effects of volcanic eruptions on planetary temperature, precipitation, and wind patterns. This research will help policymakers better understand the climate risks associated with volcanic eruptions and develop appropriate adaptation strategies.

Improving volcanic risk management:

Managing volcanic risks involves mapping risk areas, planning urban zones, raising awareness among populations, and implementing alert systems and evacuation plans. Scientists must work closely with local authorities to develop effective strategies for volcano risk management, using the latest scientific knowledge to assess risks and recommend appropriate actions.

Understanding the impact of volcanoes on the environment and ecosystems:

Volcanoes have a significant impact on the surrounding environment and ecosystems through lava flows, ash, volcanic gases, and lahars. Scientists must continue studying the impact of volcanic eruptions on soils, rivers, lakes, and oceans, as well as biodiversity and surrounding ecosystems. This research will help policymakers develop appropriate conservation and restoration strategies to protect endangered species and vulnerable habitats.

Encouraging research on extraterrestrial volcanoes:

While research on terrestrial volcanoes has made significant progress, our understanding of extraterrestrial volcanoes remains limited. Scientists must encourage research on volcanism on other planets, moons, and celestial bodies, using sophisticated measurement instruments to study the characteristics of extraterrestrial volcanic eruptions. This research will help us better understand the geological processes that have shaped our solar system and the possibilities of habitability on other celestial bodies.

In conclusion, volcanology is an exciting science that continues to reveal new challenges in understanding and predicting volcanic eruptions, protecting populations and surrounding ecosystems, and expanding our understanding of volcanic phenomena throughout our solar system

Acknowledgment

Dear reader,

We have reached the end of this journey through the mysteries and wonders of the world of volcanoes. You have been my companion throughout these pages, and I want to express my deep gratitude and a touch of nostalgia for your unwavering commitment to reading until the very end.

Throughout the chapters, we have explored the incredible history of volcanoes together, these «Guardians of the Earth» that have shaped our planet and continue to do so. We have marveled at the titanic forces that stir our world, the extraordinary processes that give rise to these giants of fire, and the impacts they have on our environment and societies.

The study of volcanoes is an endless quest, an enthralling odyssey that pushes us to constantly push the boundaries of our understanding. It's as if we were children, filled with wonder and curiosity, discovering for the first time the hidden treasures beneath the surface of our mother planet.

While writing this book, I drew from numerous reliable sources and cross-referenced information to provide you with an enriching and captivating read. I have endeavored to present you with a comprehensive panorama of the world of volcanoes, infused with the passion that drives me in this fascinating field. My deepest wish is that this reading has ignited in you the same flame of enthusiasm.

Like the eruption of a volcano, the creation of this book has been a true eruption of ideas, research, emotions, and memories. Memories of my own experiences at the feet of dormant giants, of encounters with fellow enthusiasts, and of those magical moments when the power of nature reminds us of how small and vulnerable we are in its presence.

Along the way, I have constantly marveled at the beauty and complexity of our world. I wanted to share this emotion with you, through simple language and colorful anecdotes.

As you close this book, I hope that I have imparted to you some of my love for volcanoes and offered you a new perspective on these magnificent «Guardians of the Earth». May you, in turn, share this passion and curiosity with others as we have journeyed together through these pages.

Thank you once again for sharing this voyage with me.

Yours sincerely,

Printed in Great Britain
by Amazon